D0893390

FINE POWDERS

Preparation, Properties and Uses

Advisory Editors

LESLIE HOLLIDAY,

*Visiting Professor, Brunel University,
Uxbridge, Middlesex, Great Britain*

A. KELLY, Sc.D.,

*Deputy Director, National Physical Laboratory,
Teddington, Middlesex, Great Britain*

FINE POWDERS
Preparation, Properties and Uses

C. R. VEALE

Division of Inorganic and Metallic Structure
National Physical Laboratory, Teddington, Middlesex

HALSTED PRESS DIVISION
JOHN WILEY & SONS, INC.
NEW YORK

Published in the United States and Canada by
Halsted Press Division
John Wiley & Sons, Inc., New York

Library of Congress Cataloging in Publication Data
Veale, C. R.
                    Fine Powders

"A Halsted Press book."
Includes bibliographies.
1. Powders. I. Title.
TA418.78.V4   620'.43   72–5210
ISBN 0–470–90489–5

WITH 17 TABLES AND 25 ILLUSTRATIONS

Printed in Great Britain by Galliard Limited, Great Yarmouth, Norfolk, England.

CONTENTS

FOREWORD

In the Division of Inorganic and Metallic Structure and its predecessors in the National Physical Laboratory, work on the use of plasma as a reaction medium has been in progress for several years. Because of the very steep temperature gradient at the edge of the plasma, solid products usually condense as fine powders. The Division has thus extensive expertise in the preparation of fine powders by condensation processes and considerable experience of their characterisation and of some of their properties. A number of industrial contacts and contracts has led to an appreciation of the current uses of fine powders.

It is this experience which has been drawn upon in the preparation of this short monograph in which an attempt is made to review the methods of preparation and characterisation of fine powders, together with their distinctive properties and present uses. While excellent texts exist covering in detail some of these separate aspects, it has become apparent, while working with, and thinking about, fine powders, that there is no simplified and unifying review available.

Because of the exceedingly wide area to be covered and the technologically advanced state of some of the industries concerned, the coverage is broad rather than deep, but it is believed that this is the better course in order to keep within a limited compass. In terms of the science, there is much to be learned in many of the applications of fine powders, and perhaps some reader will be encouraged to extend our meagre knowledge in these directions. Further, it is hoped that some unifying themes emerge from the consideration of the properties and uses of fine powders, a very wide variety of which may now be prepared as much more than laboratory curiosities.

CHAPTER 1

INTRODUCTION

Many centuries ago ancient man found that clay was a good con-
structional material; probably earlier still he began to use pigments
and dyes from mineral and vegetable sources to adorn himself
and the walls of his cave. In ignorance, and in the crudest way,
man had begun to use solid particles of colloidal dimensions. Much
has happened since that time. Today, on the one hand the use of
catalysts of high surface area forms the basis of whole industries;
on the other hand, we are concerned about pollution of the atmosphere by dusts and fumes containing fine particles.

Finely-divided solids, or those with large surface area, are thus
of considerable importance and long standing, but it is not considerations of this nature that have led to the present survey of fine powders.
As technology advances, chemical and physical processes are carried
out at higher and higher temperatures. Methods of generating and
containing these temperatures tend to lead to very steep temperature
gradients at the edges of hot zones. This can be seen in arcs, flames
and plasma. At these very high temperatures all reactants and
products are gaseous but cool rapidly at the edge of the hot zone,
often in conditions of very high supersaturation, and solid products
are obtained as unusually fine powders. Such powders may contain
particles of sizes down, almost, to the effective limit of the electron
microscope, and may have surface areas of several hundreds of
square metres per gram.

In the present context, the term 'fine powder' is used to cover
materials in which either the ultimate particle size is less than 1 μm
or the equivalent sphere diameter calculated from the surface area
is less than this value. This means, in fact, that surface areas are
greater than about 1 m^2/g. In such materials a relatively high

proportion of the atoms are on, or close to, the surface and there is a high density of potential bonding sites. It will be shown later that it is the extent and nature of this surface which gives fine powders many of their distinctive properties and applications.

Solids of high surface area are much more than an academic curiosity. Table 1 gives an idea of the UK production and cost of the three materials most used in this form. It is evident that these are commercial chemicals of some importance and, indeed, of fairly long standing since carbon black was first made more than seventy years ago.

While, with the exception of carbon black, single oxides are the most numerous and important group of materials which have been made as submicrometre powders, current studies suggest that a

TABLE 1

PRINCIPAL FINE POWDER MATERIALS[1]

	UK production 1968 (tons)	Cost/lb
Carbon black	181 000	from about 4p
Silica	>10 000	from about 5p
Titania	110 000	about 9–12p

much wider range of materials, including metals, carbides, nitrides, etc., could now be prepared if desired. The present survey was therefore undertaken to obtain an overall view of the methods of preparation, characterisation, properties and current uses of fine powders. In the course of this survey it has been necessary to impinge somewhat on the fields of smokes and of aerosols, both of which may be regarded as dilute suspensions of fine particles.

Some of the current uses of fine powders are in industries which are highly developed technologically, but in which the very complexity of the operations makes it difficult to establish the science. For example, in the reinforcement of elastomers many additives are used for various purposes and it is difficult to understand their interactions. As a result, the survey of the applications of fine powders may, at times, appear a trifle superficial, but it is hoped that it will be sufficiently comprehensive to be of value in stimulating new applications and extensions of older ones. Some topics receive much less than a fair mention in this survey. Fine powders of organic compounds are not mentioned; this is an omission reflecting current interests. The preparation, structure and uses of carbon black are treated very briefly; to do otherwise would have necessitated a separate survey of this one subject. The same is true of the field of catalysis which contains much art and does not permit of satisfactory generalisations.

The first two main chapters follow a fairly logical order in considering, respectively, methods of preparation of fine powders and methods of characterising them.

The chemical and physical nature of the surface of a fine powder is of paramount importance whenever such a powder is set in a matrix. In Chapter 4 we look briefly at the surface chemistry and surface modification of one or two of the materials of commercial importance in fine powder form. We are, in the main, concerned with major changes in the nature of the surface rather than with the more subtle and detailed changes involved in physical adsorption and chemisorption which are of interest in the study of catalysis. Subsequently an attempt is made, in Chapters 5 and 6, to see how far current applications take account of possible variations in the surface of the powder. While the influence of the extent of the surface is widely appreciated and exerts a major influence on properties in composite systems, it seems that the chemistry of the surface has hitherto been of less concern. This situation is now changing.

Consolidation of a powder involves interaction between the surface of adjacent particles and may also be susceptible to changes in the surface itself. Consolidation—sintering and/or hot-pressing—is considered in Chapter 7. The main effect of compacting fine powders is likely to be the possible preservation of a fine-grained microstructure in the pressed material, and it is therefore necessary to define the useful properties of such a microstructure.

The remaining properties of fine powders are dealt with in Chapter 8 which concludes with a short section on gas–solid reactions relevant to the handling of fine powders and their stability. While it is possible to devise conditions under which fine particles of metals and reactive compounds may be handled in safety, because of the additional cost involved, reasons for using these compounds must be most cogent.

Where it is reasonable to do so, suggestions are made as to possible lines of future development of fine powders, and the general theme which emerges is recalled in Chapter 9.

REFERENCES

1. Sources: *Annual digest of statistics*, HMSO London (1971). *Chemical Age*, (1971), **102**, (2686), 17. *ibid* (1970), **101**, (2681), 10. *Paint Oil and Colour J.*, (1971), **160**, (3801), 281.

CHAPTER 2

PREPARATION OF FINE POWDERS

INTRODUCTION

The production of fine powders, as defined in Chapter 1, may be accomplished either by breaking-down or by building-up. Whether breaking down coarser particles by grinding or building up by growth of finer particles, the process must be stopped at the required point. It is not often possible to control particle growth so as to give a fine powder of closely-controlled uniform particle size.

BREAKING-DOWN PROCESSES

Breaking-down processes are very widely applicable. Methods may be classified as shown in Fig. 1. Atomisation occurs when a jet of liquid is converted to very small droplets, for example by impinging the liquid jet onto a rapidly rotating disc or forcing it through a small orifice under pressure. The frozen droplets are predominantly larger than the particles under consideration here; only a small proportion of sub-sieve sized material being produced.[1]

Grinding in rod mills, ball mills or other types of mill is very

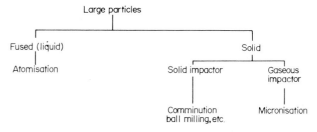

Fig. 1.   *Breaking-down processes.*

5

widely used in the size reduction of rock, coal, etc. The mechanics and energetics of the process have been reviewed.[2,3] In the present state of the art theories of particle fracture and comminution do not have much significance as it is not possible to correlate theory and practice because some of the required experimental parameters cannot be measured. Although the end product of grinding is usually relatively coarse, fine powders can be obtained. At some stage in the process re-aggregation (cold welding or agglomeration)

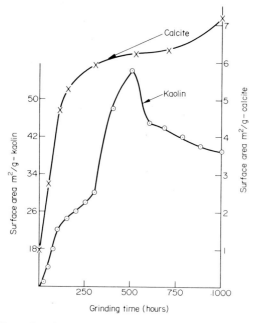

Fig. 2.   *Effect of prolonged grinding on the surface area of some powders.*[4]

may become as important as size reduction and a rough state of equilibrium may be reached. Examples of this are shown in Fig. 2.[4] On the other hand, it is possible that the grinding limit is apparent rather than real in that size reduction becomes slower and it becomes more difficult to retain the smaller particles. This is more likely to occur with hard and brittle materials which would not readily self weld. The addition of so-called 'grinding aids' to the grinding fluid serves to increase the efficiency of the process. These grinding aids may be surface-active agents or inorganic salts containing multivalent ions. If the grinding aid is adsorbed on the surface of the material being ground, the surface free energy of the latter will be reduced and, for a brittle material, its strength—as indicated by the well-known Griffith equation—will be reduced. This supposition is, to some extent, supported by the fact that the nature of the grinding fluid and the polarity of the multivalent inorganic ion both exert but little effect on the grinding of some metal powders.[5]

The stresses to which particles are subjected during grinding have been estimated at hundreds of MN/m^2.[6] Where the area of strained surface is large (fine powders) the effect of grinding may be noticeable. Ground quartz has a surface layer about 100 atoms thick which is amorphous to X-rays and shows increased physical and chemical reactivity.[7] The presence of 'broken bonds' on the surface of several freshly and finely ground materials has been considered to be the cause of an increased intensity of electron spin resonance signals.[8] Owing to chemisorption, and consequent removal of free electrons, these signals subsequently decay.

Micronisation depends solely on particle–particle contact for size reduction. Particles are thrown together in a turbulent air or steam jet. The overall spiral path followed by the material leads to the ejection of coarser particles[9] and to a size distribution with a sharp upper cut off but poorly-controlled distribution. Micronisation is now used quite extensively to provide pigment particles, usually of a few micrometres diameter, but, sometimes, as with titanium dioxide, of submicrometre sizes.

BUILDING-UP PROCESSES

In order to prepare a fine powder by a process of building-up there must be a phase change. There may also be an overall chemical reaction, but this is not essential. In principle, any type of phase change which results in a solid product may be suitably controlled to yield a fine powder. Changes which have been so used are shown in Table 2. Since the processes of nucleation and growth proceed almost simultaneously, there are often difficulties in producing the required large number of nuclei and preventing them from growing too far.

For details of theories of nucleation and growth any of the standard texts may be consulted, e.g.[10,11] Here it will be sufficient merely to consider the factors which control these phenomena.

TABLE 2

PHASE CHANGES WHICH HAVE YIELDED FINE POWDERS

Phase change	Process yielding fine powder
Vapour–liquid–solid	Condensation with liquid droplets subsequently solidifying
Vapour–solid	Direct condensation to a solid
Liquid–solid	Crystallisation from solution, often as precipitation
Solid–solid	Unmixing of solid solution. Solid phase change. Decomposition solid $\rightarrow$ solid + gas

Nucleation

The formation of a nucleus—a small volume element of a new phase—can occur in two ways. It may occur homogeneously as a result of the spontaneous growth of clusters of the same composition, or heterogeneously as a result of catalytic action by foreign particles, the walls of the vessel, etc., when the initiating agent will be of different composition from the rest of the nucleus. Heterogeneous nucleation requires the lower degree of supersaturation and thus predominates when the conditions approach equilibrium. When conditions are very far removed from equilibrium and nucleation is rapid, homogeneous nucleation predominates.

The formation of a nucleus involves the creation of a boundary. Although the free energy of the system will be reduced by the formation of the new phase, the newly-created boundary will have associated with it a finite amount of free energy. If the overall result is an increase in free energy, the nucleus is unstable. At a given temperature the overall free energy change is related to the radius of the nucleus in the manner shown in Fig. 3. The classical theory of Becker and Döring applied macroscopic thermodynamic concepts to such a microscopic system as a nucleus, in order to obtain an expression for the rate of homogeneous nucleation at relatively low supersaturation. To overcome the weakness inherent in this, approaches using a statistical mechanical treatment of the thermodynamics of small clusters[12] and of inhomogeneous systems[13] have been applied.

In general, the accepted form of the expression for the rate of nucleation consists of the product of two exponential factors. Of these, one is an expression of the probability of forming a nucleus of greater than the critical radius, and the other is concerned with the rate at which material can be supplied to the nucleus in the

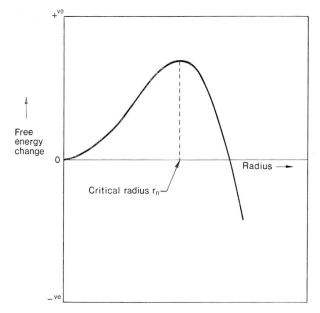

Fig. 3. Change in overall free energy as a nucleus grows.

required structure for growth. In nucleation from the gas phase, this second exponential can be ignored, but where nucleation is taking place from the liquid or, more especially, from the solid, this factor soon becomes controlling. In the exponential expressing the probability of forming a supercritical nucleus there is a dependence on p/p_e the degree of supersaturation. The increase in the rate of nucleation as p/p_e increases is exceedingly rapid, as can be seen from Fig. 4.

When the degree of supersaturation is high, the system departs radically from equilibrium and another limitation of classical nucleation theory is encountered. In a relatively recent approach to such conditions[14] equilibrium concentrations of critical and sub-critical nuclei are expressed as a partition function of clusters, and coagulation of sub-critical clusters is included. Application of this model to silver powder produced by rapid chilling of a hot saturated argon stream showed promising agreement between experimental surface area and calculated size distribution.

Growth of a solid embryo is further complicated by the need to consider the crystalline growth parameters mentioned below.[10,11]

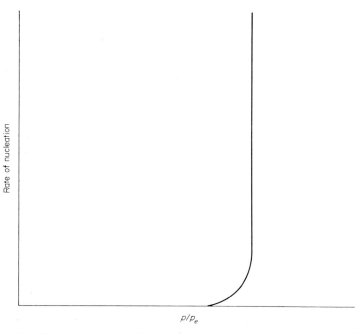

Fig. 4.   Change in the rate of nucleation with increasing supersaturation.[16]

The theory of heterogeneous nucleation on foreign particles or surfaces has been considered but is much less amenable to theoretical treatment. A modifying term is usually applied to the expression for homogeneous nucleation and provided that the limitation in the number of foreign nuclei is borne in mind, similar considerations apply.[10,11]

Growth

Once a nucleus has exceeded the critical size it will continue to grow at a rate controlled by the provision of fresh material in the correct orientation at the surface of the growing nucleus and by the ease of formation of new bonds in the required positions. When a crystal is growing, the great complexity of the growth process must be recognised, together with the strong influence of crystal dislocations of all sorts, especially spiral dislocations. Repeated nucleation of new crystal planes is not necessary. The subject has been considered in some detail.[10,11,15]

With each of the types of phase change listed in Table 2 there are different aspects of the nucleation and growth processes which are of especial importance, but the general approach is the same.[10,11,15,16,17]

Application to the production of a fine powder

The requirement for the production of a fine powder is initially for the formation of a large number of nuclei, the subsequent growth of which is restricted. To produce many nuclei a high degree of supersaturation is necessary and, as a result, nucleation will tend to be homogeneous. In a condensation a high degree of supersaturation may be achieved by supercooling. Under these conditions the size of the critical nucleus is approximately inversely proportional to the degree of supersaturation. A bulk of one hundred molecules has been indicated for water formed from the supercooled vapour at $300°K$[11] and work with the field-emission microscope has suggested somewhat similar sizes for other materials.[18]

In some other types of phase change (*e.g.* precipitation or thermal decomposition) conditions are different and supersaturation is produced by exceeding the solubility product or merely by continued heating in the case of a thermal decomposition.

Having produced multiple nucleation, growth must be restricted. In condensation from the gas phase, growth is very rapid and can be controlled only by the removal of the fine powder and the provision of very large numbers of nuclei. Both these criteria are fulfilled by very rapid quenching. Growth from a liquid, and, even more, growth from a solid, is slower because of diffusion control and the need to present the diffusing material to the growing particle in the correct orientation. Control of growth is consequently less difficult in these instances.

The practical difficulties of preparing a fine powder depend considerably on the scale of operation. Colloidal suspensions and aerosols cannot strictly be termed fine powders, but they contain fine particles: however, the amount of material available for particle growth may be very limited and growth is consequently less difficult to control. In an aerosol generator[19] nucleation is probably heterogeneous under normal conditions, but is often made so purposely by the introduction of foreign nuclei obtained, in their turn, by evaporation and unseeded condensation. This gives rise to a more reproducible aerosol with a narrower size range. For example, particles of sodium chloride in the size range 150–1 000 nm were obtained by uncontrolled condensation from a current of helium or argon.[20] However, by injection of sodium fluoride nuclei into the gas stream, the sodium chloride size fell in the range 200–400 nm.[21] The selection of a nucleating material is governed by its volatility relative to that of the aerosol material.[22] Crystal habit can be modified by slight changes in the composition of the gas stream. Thus, when a number of metals were evaporated into a low pressure (10 torr) current of argon, the presence of about 0·5 per cent v/v of oxygen caused changes in the shape of the crystals which became more irregular and roughened,[23] apparently without gross change in their chemical composition.

Now that atmospheric pollution by fine dusts is a source of concern, the study of aerosols is of widespread interest and new methods for their preparation are constantly being devised[24] and conditions for their coagulation delineated.[25] Although the aerosol route may be used to prepare a very wide variety of fine particles, the rate of production is exceedingly slow; 0·5 g/month has been quoted.[21]

A very wide range of sols or colloidal suspensions of solids in liquids can be prepared by precipitation in very dilute solutions, but, again, the amount of solid produced is usually small. However, in suitable instances precipitation can lead to macroproduction of a fine powder (*see* page 21).

It will now be appropriate to consider the macroscale production of fine powders from each of the phase changes and to give a few examples showing the types of powder prepared using each change.

Evaporation–condensation without overall chemical reaction

Oxides, metals and carbides are among the materials which have been distilled. For these, high temperatures are required, but rapid cooling is more easily achieved.

In some instances, a result of this rapid cooling is the retention of a metastable form of the product. This occurs, for example, with alumina where the δ-form is usually obtained rather than the α-form which is stable above about 1 000°C. It has been suggested that the explanation is that a higher temperature increases the tetrahedral (as against octahedral) coordination of aluminium. In liquid alumina tetrahedral coordination predominates[26,27] and it is possible that this coordination is retained on quenching, leading to δ-alumina, while, under equilibrium conditions, there is time for the octahedral coordination which occurs in α-alumina to be taken up. An alternative explanation is that nucleation of solid may occur only at a degree of supercooling such that the γ- or δ-form of alumina is the stable phase, while the heat of solidification may be sufficient to raise the material above the transition temperature only for larger particles.

In order to achieve the high temperatures required, one popular heat source which has the advantage of being usable in controlled atmospheres, is the high intensity arc.[28] In this the feed material, mixed with graphite if itself non-conducting, is made the anode. Temperatures may reach 7 000°K in the tail flame as the vapour jet streams from the anode. Once out of the arc flame, the temperature of the vapour falls rapidly and the resulting high degree of super-saturation leads to rapid condensation. The use of the high intensity arc in the production of a wide variety of fine powders has been reported[29] and many of these have been available commercially.[30] Some of these powders are the result of overall chemical reaction (*see* next section), but this is not so for oxides evaporated in air

TABLE 3

SOME FINE POWDERS PREPARED USING THE HIGH INTENSITY ARC[29,30]

(A) Single components			
Oxides (in air)			
SiO_2	Al_2O_3	Fe_2O_3	ThO_2
MnO_2	Nb_2O_5	NiO	Y_2O_3
UO_2	MoO_3	ZrO_2	MgO
WO_3			
Elements (in an inert atmosphere)			
C	Al	Li	Ni
Fe	W	Mo	
Carbides (in an inert atmosphere)			
ThC	TiC	B_4C	UC
TaC	SiC		

(B) Polycomponents		
Oxides		
(Zn, Mn, Cu, Fe)O	$Al_2O_3 . SiO_2 . Fe_2O_3$	
$LiAl(SiO_3)_2$	$MnO . SiO_2$	
$FeO . Cr_2O_3$		
Carbides		
$B_4C . SiC$	$UC . NbC$	
$UC . ThC$		

Fig. 5.   High intensity arc in operation.

or metals or carbides evaporated in an inert atmosphere. Some idea of the range of materials so obtained is given in Table 3. A high intensity arc in operation is shown in Fig. 5.

Mean particle diameters usually lie in the range 5–100 nm. With most materials the particles are spheres. Angular particles are thought to result from vapour–solid condensation, while spheres are formed when there is a liquid intermediate.

One of the most efficient ways of using plasma heating is the centrifugal liquid wall furnace. For many purposes, for example when it is necessary to prepare hundreds of grams of material, this method is more convenient than the use of the high intensity arc. The material to be volatilised is cast in the form of a hollow cylinder. This is then made the core of a furnace and the plasma flame plays down the central hole in the core, fusing and vaporising the inner surface of the ceramic. The molten refractory forms a stable layer

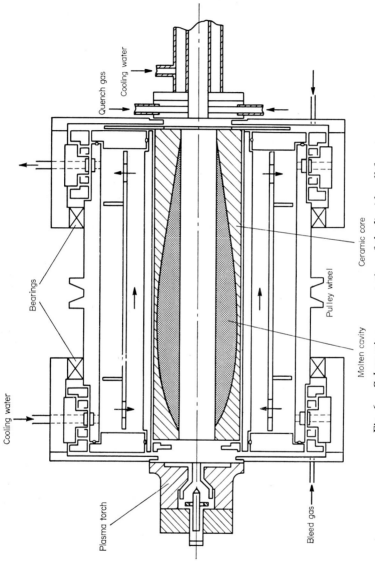

Fig. 6. Schematic representation of the liquid wall furnace.

Fig. 7.   The centrifugal liquid wall furnace.

held in place by the centrifugal action of the rapidly rotating furnace. This is shown diagrammatically in Fig. 6, while Fig. 7 shows a furnace in its operating environment. In this way, as in the high intensity arc, the difficulty of injecting powder into the relatively high density plasma stream is avoided. The centrifugal liquid wall furnace also permits a degree of control to be exercised over the condensation conditions.[31,32] Thus, the so-called 'active' silicas are condensed in the presence of hydrogen-containing compounds. At the highest magnification, electron micrographs of these compounds show the existence of a web-like structure between the spherical or angular particles.[33] By quenching with a non-reactive gas an 'inactive'

TABLE 4

SOME FINE POWDERS PREPARED USING THE CENTRIFUGAL LIQUID WALL FURNACE[31,32]

Oxides		
SiO	SiO_2	Al_2O_3
TiO_2	MgO	

(*i.e.* non-thixotropic) silica is obtained.[34] Some examples of fine powders which have been prepared with the centrifugal liquid wall furnace are given in Table 4.

Evaporation with condensation after chemical reaction

Chemical reaction may occur almost incidentally in the course of the evaporation–condensation process, or it may be necessary to make special provisions for such a reaction. Examples of the first case are to be found in the formation of fine metal oxides by striking an arc in air using the metal or alloy as electrodes,[35] or in the use of the high intensity arc in hydrogen atmosphere to prepare fine nickel, iron or tungsten from the corresponding oxides.[29]

Examples of special provisions are to be found in the control of the degree of surface hydroxylation of fine powders by control of the water content of the atmosphere in which condensation occurs[31,32] or in the provision of a specific reactive atmosphere in which vapour–vapour reaction may take place. By means of this last procedure an exceedingly wide range of materials such as metals, metalloids, oxides, carbides, borides, nitrides, silicides, sulphides and many others, can be prepared.[36]

Other reactions falling into this category may be of the type:

$$\text{vapour} + \text{vapour} \rightarrow \text{solid} + \text{vapour}$$

Depending on the reaction conditions (temperature, rate, concentration of reactants, condensation conditions, etc.) the product can, on occasion, take the form of an epitaxial film, platelets, whiskers, polycrystals, amorphous deposit or fine powder. Once more, to produce a fine powder, multiple nucleation and restricted particle growth are necessary. The latter may entail quenching, or it may be that conditions of concentration, gas flow, etc. and the rate of provision of nuclei are such that particle growth is naturally restricted. An example of such natural control of growth is the hydrolytic decomposition of volatile alkoxides of titanium, zirconium, hafnium, etc. by a moist current of inert gas.[37]

The heat source should be appropriate to the temperature required. For high temperature reactions, arcs, chemical flames,[38] direct[39] or induction plasmas[40] may be used. Of these, the latter, having no electrodes, may be operated in either corrosive or highly-oxidising environments, and with a range of different torch designs. The addition of facilities for the admission of quench gas results in a

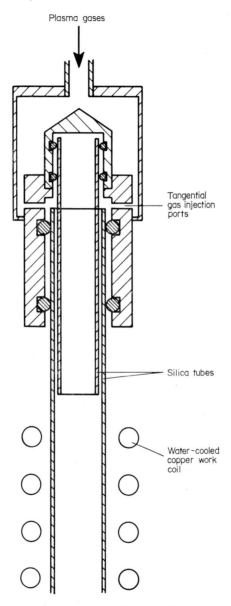

Fig. 8. *Schematic representation of a radio-frequency induced plasma device.*

highly flexible assembly capable of exerting close control over all facets of any vapour–vapour reaction and of working over a very wide range of vapour concentrations.[41] Equipment for using the induction plasma is shown diagrammatically in Fig. 8, and Fig. 9

Fig. 9.   RF induced plasma in operation.

TABLE 5

EXAMPLES OF SOME VAPOUR PHASE REACTIONS WHICH HAVE PRODUCED SUBMICROMETRE POWDERS

Vapour phase reaction	Temperature	Materials prepared	References
Oxidation of chloride or oxychloride by NO_2	~400°C	Nb_2O_5 MoO_3 WO_3 B_2O_3 V_2O_5	42
	175–500°C	Al_2O_3 + aluminium oxychloride	44
Oxidation of chloride by oxygen	1 000–1 700°C	TiO_2 Al_2O_3 ZrO_2 SiO_2 ZnO Cr_2O_3 Fe_2O_3	45, 46, 48
Oxidation of chloride or oxychloride by oxygen	(>5 000°K) (plasma)	αCr_2O_3 δAl_2O_3 Cr_2O_3–Al_2O_3 solid solutions TiO_2 (anatase)	40
Reduction of chloride by hydrogen	800°C	Mo W	38, 39, 52
Reduction of chloride by hydrogen in methane or in chlorinated hydrocarbon	~3 000°C	TaC TiC NbC SiC (plasma)	39, 56
	plasma	TiC	55
Reaction of volatile oxide with ammonia in nitrogen		BN	54
Reaction of volatile halide with ammonia	1 500–2 000°C	AlN Si_3N_4 BN Zr_3N_4 TiN	57
Reduction of chloride by hydrogen in nitrogen	~3 000°C	TaN	39
Reduction of volatile fluoride by hydrogen	H_2–F_2 flame	W Mo W–Mo Alloy W–Re Alloy	47
Reduction of silica by carbon	arc	SiC	50
	~1 400°C	SiC	
Hydrolysis of volatile metal halides	flame	SiO_2 Al_2O_3	51, 53
Thermal decomposition of metal alkoxide vapours	320–450°C	TiO_2 ZrO_2 HfO_2 ThO_2 Y_2O_3 Dy_2O_3 Yb_2O_3	37
Oxidation of metal alkyl by burning		Al_2O_3	49
Reduction of oxide by methane	plasma	SiC	58

shows a plasma in operation. Some examples of the various reactions which have yielded fine powders are given in Table 5.

Many of the fine powders produced on an industrial scale result from vapour phase reactions. Such materials are carbon blacks for use in reinforcing elastomers and as a pigment, titanium dioxide and, to a lesser extent, zinc and antimony oxides for use as pigments and silica as a filler and reinforcing agent. It is not proposed to make a detailed survey of these processes, which are basically of three types:

(1) oxidation or hydrolysis of volatile chlorides (SiO_2, TiO_2)[42];
(2) oxidation of metal and formation of oxide smoke (ZnO, Sb_2O_5);
(3) thermal decomposition and/or dehydrogenation of hydrocarbons, giving carbon blacks.[43]

Each process has an extensive patent literature.

Precipitation

With this process, once again, in order to produce a fine powder, a large number of nuclei must be formed and growth restricted to keep the particles small. Rates of nucleation and growth vary widely in reactions in which a precipitate is formed and, consequently, there are wide differences in the effect of conditions (for example concentration or degree of mixing) on the particle size of the final precipitate.

After precipitation is complete there is still change in the particle size distribution in the precipitate. There are at least two causes for this. One is the greatest solubility of the finer particles—the so-called 'Ostwald ripening'. Because of this, the coarser particles grow at the expense of the finer.[60] In addition, normal recrystallisation occurs and tends to 'cement together' the precipitate particles.[61] The extent to which this latter event occurs is dependent on the absolute solubility since it involves the deposition of material at necks between particles as a result of exchange between the surface of the solid and the solution. Absolute solubility, in its turn, is affected by change of pH, presence of common ions or neutral salts, etc. To some extent, therefore, the size of precipitated particles is also dependent on these factors. This has been demonstrated for zinc sulphide precipitates.[62]

In addition to growth of the actual primary particles of a precipitate, flocculation usually (but not always) also occurs. This loose binding together of the primary particles encloses greater or lesser amounts of liquid phase and does not, in itself, result in a major loss of available surface area. The surface area of the flocculated precipitate is largely dependent on the size of the individual particles which, in turn, is dependent on the ionic concentration in the solution at the time of precipitation. As an example, the surface area of a titania gel decreases by a factor of 150 when the pH of the solution in which the initial precipitation takes place is lowered from 4 to 1, with a consequent increase in the ionic content of the solution.[63]

The rapidity with which flocculation occurs in the presence of electrolytes varies with the material. The so-called lyophobic sols are very unstable (*e.g.* arsenious sulphide and silver chloride) while the lyophilic sols, such as hydrated oxides, are much more stable. Indeed, if the electrolyte concentration is reduced by some means such as dialysis, solvent extraction, ion exchange or washing, stable sols of 6 M concentration of some hydrated oxides may be prepared.[64] These sols may be further concentrated or converted to coherent forms, either by removal of water by evaporation or solvent extraction, or by removal of the anion stabilising the sol,[64] or by the addition of gelling agents such as dextran.[65] Gels retain the fine particle size and this can be utilised in sintering or other processes.

An example of a reaction which, with changes in the conditions, can produce a stable fine powder or a gel, is the hydrolysis of metal alkoxides to form hydrated oxides. The gels, formed when hydrolysis is not immediately complete, may be dried either under supercritical conditions in an autoclave (aerogel) or in vacuum (xerogel), are very fluffy, and contain considerable internal porosity.[66] The powders are formed when hydrolysis is immediate and complete. Several variations have been used with hydrolysis by a current of steam or by the addition of liquid water or by adding the alkoxide to a slight excess of water. In all instances, the reaction mixture is vigorously stirred.[67] As far as can be detected, the more rapid reaction (in order to form a fine powder) is the only relevant difference between these two procedures. One of the important advantages of precipitation processes for the preparation of fine powders lies in their ability to prepare compound double oxides following simultaneous precipitation of the component hydrated oxides, oxalates, etc. However, it is possible that the preparative conditions (*e.g.* pH, concentration, rate of mixing, adsorption) may preclude quantitative stoichiometric and simultaneous precipitation and prevent the formation of compounds and solid solutions or mixtures.

Much of the silica used industrially (for example, as a filler in silicone rubber) is prepared by precipitation processes. A solution of sodium silicate is neutralised in one of a wide variety of ways with the process variables manipulated to prevent excessive particle growth (*e.g.* by the presence of a soluble calcium salt). Such details as are available are contained in the patent literature. The hydrated silica is then recovered and dried. Hydrated, finely-divided silicates of calcium, aluminium, etc., are also prepared by precipitation and are used for similar purposes.[68] As a variation on this type of process, it has been found possible to prepare essentially monodisperse silica spheres larger than about 0·05 μm by hydrolysis of alkyl silicates and subsequent condensation of silicic acid in alcoholic solutions.[69]

Titanium dioxide is also prepared industrially by precipitation although the vapour phase 'chloride' route mentioned above is tending now to take over. In this precipitation a solution of titanyl sulphate is hydrolysed at elevated temperatures. Seeding is employed, primarily to obtain the required rutile form of the oxide. The precipitate is initially amorphous, but crystallises on ageing. Important factors in the precipitation include the quantity of seed, the acidity and concentration of the solution and the rate of heating. Sometimes, conversion to the rutile form is produced by subsequent calcination (750–1 000°C).[70]

In many instances, the precipitate is not itself the desired product and is thermally decomposed without the particle size being determined. The finely-divided nature of the product may then result from either the decomposition or the precipitation.[71] However, some catalysts, prepared by precipitation, are used without further treatment. Thus, precious metal catalysts are prepared by reduction and precipitated on suitable substrates. One of the functions of the substrate in such instances is to prevent sintering of the powder at the temperature of use by keeping the particles out of contact with each other. Such a material is platinised asbestos, in which crystallites of platinum adhere to the fibres of the asbestos.

It is also possible to obtain a powder, the particles of which fall within the adopted definition of 'fine' without using a support, but the available surface area is considerably reduced. Submicrometre particles of nickel may be obtained by reduction, with hydrogen, of a nickel amine complex under pressure.[72]

TABLE 6
SOME FINE POWDERS
OBTAINED BY
PRECIPITATION

Compounds	References
$CoFe_2O_4$	73
$NiFe_2O_4$	74
$CuFe_2O_4$	74
$BaTiO_3$	75
$MgAl_2O_4$	78

Solid solutions	References
$BaTiO_3$–$SrTiO_3$	75
Al_2O_3–TiO_2	76
Al_2O_3–SiO_2	77

Some examples of compounds and solid solutions obtained in fine powder form by precipitation are shown in Table 6.

Thermal decomposition

The theory of the nucleation and growth processes for a thermal decomposition (specifically of salt hydrate crystals) has been considered.[16] If the thermal decomposition takes the form:

$$A_{solid} \rightarrow B_{solid} + C_{gas}$$

the formation of the new phase B results from local structural fluctuations in the lattice of A which produce conditions favourable for the formation of a nucleus of B. Such sites are situated at regions of disorder *e.g.* vacancy, interstitial or impurity clusters, meeting points of high angle grain boundaries, etc.

In most decompositions the molecular volume of the product is less than that of the reactants. Because of this volume change, both reactant and product crystals become strained and this strain cannot be relieved until the critical shear stress in the neighbourhood of the nucleus is exceeded. Thus, there is a strain energy involved in the growth of a nucleus up to, and especially beyond, its critical radius. Typically, the interfacial strain energy is of the order of 1 kcal/mole, and this is sufficient to account for the slow growth of small nuclei which leads to the formation of a fine powder.[79]

For a particular chemical decomposition the rate of formation of nuclei is governed, in part, by imperfections in the crystal lattice of reactant A. In some instances the stereochemical configuration at the imperfections seems to be important. For example, in the decomposition of calcium carbonate the reactivity of the solid is enhanced only at certain dislocations.[80] Growth, in turn, may vary with the particle size and method of preparation of reactant A. Thus, the formation of anhydrous barium styphnate from the monohydrate is much more rapid when the monohydrate is in the form of small crystals than when the monohydrate crystals are large. This is due to surface nucleation followed by preferential growth along the surface or grain boundaries (where strain energy is less) prior to penetration of the crystallites.[81] Similarly, the thermal decomposition of freeze dried (high surface area) ferrous sulphate is faster than that of the reagent grade material under comparable conditions.

Study of a large number of decompositions of oxycompounds shows that topotactic behaviour is very common and may be customary, so that the oxide products show a definite orientational relationship to the starting material. Thus, magnesia prepared from needles of magnesium carbonate trihydrate retains the needle shape in spite of about 70 per cent reduction in volume.[83] Again, uranium dioxide has been prepared by thermal decomposition of nine different materials, each with its own macroscopic shape. The result has been nine different pseudomorphs of uranium dioxide.[84] This does not mean that, in order to prepare a fine powder (*i.e.* a powder with large surface area) by thermal decomposition, the starting material must be in fine powder form. The macroscopic form of the product is no indication of its microstructure. Decomposition of magnesium hydroxide in either massive form or as submicrometre hexagonal platelets produces magnesia of the same order of surface area but resembling the starting material in visible shape.[85] The surface area of the product depends on the density of nucleation inside each grain of reactant. Each nucleus grows, and when the material is unable to withstand the shear strain resulting from the volume change on decomposition, splitting and/or channelling occur, forming suitable avenues for the escape of the volatile reaction product and so assisting the reaction to proceed. The particle size of the final product is a compromise between the need for the decomposition to occur to near completion, and sintering of the product. The sensitivity of sintering behaviour to small concentrations of impurity is well known and is relevant here. For example, iron oxide grows more rapidly in the presence of water vapour than in a dry atmosphere.[82] Sintering may

TABLE 7

SOME FINELY-DIVIDED OXIDES PREPARED BY ATOMISING SOLUTIONS INTO A FLAME

	Products	Starting solution	Reference
(A) Single oxides	δ-Al_2O_3 Co_3O_4 Cr_2O_3 CuO α-Fe_2O_3 γ-Fe_2O_3 HfO_2 p-MnO_2 Mn_2O_3 NiO SnO_2 ThO_2 TiO_2 ZrO_2	Sulphate or chloride	89
	Mn_3O_4	Acetate	89
		Chlorides	89
(B) Double oxides	$CoFe_2O_4$ $MgFe_2O_4$ $MnFe_2O_4$ $ZnFe_2O_4$ $BaO6Fe_2O_3$	Acetates or lactates	89
	$BaTiO_3$	Nitrates	89, 90, 91
	(Ni, Zn) Fe_2O_4 $PbCrO_4$ $Cu_2Cr_2O_4$	Sulphates	89
	$CoAl_2O_4$	Sulphates	89
(C) Mixed oxides	Al_2O_3–Cr_2O_3 Al_2O_3–CuO Al_2O_3–Fe_2O_3 Al_2O_3–NiO Al_2O_3–NiO Cr_2O_3–Fe_2O_3 Fe_2O_3–$CoFe_2O_4$	Chlorides	89

TABLE 8

FINE POWDERS PREPARED BY REMOVAL OF WATER FROM HYDRATES

Starting material	Product	Conditions	Reference
Manganous oxalate dihydrate	Anhydrous salt	Vacuum at 60°C	92
Nickel malonate dihydrate	Anhydrous salt	Vacuum at 160°C for 18 hours	93
Copper sulphate pentahydrate	Monohydrate	<1·4 mm water vapour 40–45°C	94, 95
	Trihydrate	>1·4 mm water vapour 40–45°C	95
Cobaltous chloride hexahydrate	Monohydrate	3·5 mm water vapour 30–40°C	96, 97
Nickel sulphate tetrahydrate	Monohydrate	Vacuum at 40°C or <2 mm water vapour at 60°C	96, 97, 98
Cobalt sulphate tetrahydrate	Monohydrate	Vacuum at room temperature	98
Manganous sulphate tetrahydrate	Monohydrate	Vacuum at 40°C/5 hours or <1·2 mm water vapour at 50°C	96
Zinc sulphate heptahydrate	Monohydrate	Vacuum at 40–45°C/5 hours or <1·5 mm water vapour at 40°C	94, 96, 97
Magnesium sulphate heptahydrate	Monohydrate	Vacuum at 40–60°C or <4·5 mm water vapour at 40°C or <8 mm water vapour at 50°C or <12·5 mm water vapour at 60°C	96, 99

TABLE 9

SOME SINGLE OXIDES PREPARED BY THERMAL DECOMPOSITION OTHER THAN IN A FLAME

Product	Starting material	Conditions	Reference
NiO	Hydroxide	250–1 300°C in N_2	100
	Sulphate	1 000°C/2 hours in air	100
TiO_2	Sulphate solution	850–920°C with ~1 per cent carbon	102
		Sprayed with carbon black into fluidised bed 1 100°C	102
MgO	Hydroxide, oxalate, formate, basic carbonate	Air, 500°C	85, 103
		or 250–350°C in vacuum	83
ZnO	Hydroxide (hydrated)	120–500°C	101
Cr_2O_3	Oxalate	400°C in air	104
	Oxide gel	400–700°C	105
	Oxalate hexahydrate	330°C in air	104
ThO_2	Oxalate	650–1 300°C	106
Al_2O_3	Ammonium alum	1 000–1 100°C in air	107
	Oxalate tetrahydrate	300°C	104
Fe_2O_3	Ferric oxalate pentahydrate	300°C	104
FeO	Ferrous oxalate dihydrate	300°C	104
MnO	Oxalate dihydrate	300°C	104
CoO	Hydroxide	250–600°C	108

TABLE 10

SOME DOUBLE OR MIXED OXIDES WHICH HAVE BEEN PREPARED BY THERMAL DECOMPOSITION OTHER THAN IN A FLAME

Product	Starting material	Conditions	Reference
Nickel ferrite	Mixed oxalates	600°C in air. Incomplete reaction	111
	Mixed sulphates +~10 per cent carbon black	680°C/2 hours in air	110
Nickel zinc ferrite	Mixed oxalates	600°C in air	91
	Ferric formate, nickel oxalate and zinc carbonate	500–1 000°C in air	91
	Mixed sulphates	800–1 000°C in air	113
Nickel cobalt ferrite	Solid solutions nickel and cobalt oxalates with ferric oxalate	400°C in air	86
Lead zirconate	Lead nitrate or acetate and zirconyl nitrate	750°C in air	112
	Lead nitrate or acetate and zirconyl nitrate in aqueous oxalic acid to give double oxalates	500–700°C	112
Lead titanate	Lead nitrate or acetate and titanyl nitrate	750°C in air	112
	Lead nitrate or acetate and titanyl nitrate in aqueous oxalic acid to give double oxalates	500–700°C	112
Barium titanate	Barium titanyl oxalate	600–1 200°C in air	75
Bismuth molybdate	Precipitated mixed hydrated oxides	450°C	115
Nickel oxide—magnesium oxide solid solutions	Precipitated mixed hydrated oxides	Below 600°C in air	109
Magnesium aluminium oxide	Mixed sulphates	800°C/2 hours in air	114

TABLE 11

NON-OXIDES PREPARED BY THERMAL DECOMPOSITION

Product	Starting material	Conditions	Reference
Nickel carbide (with little nickel)	Nickel malonate	286°C in vacuum	93
Iron (with little oxide)	Ferrous sulphate $+\sim5$ per cent carbon black	820°C/1 hour in inert atmosphere	110
Aluminium carbide	Aluminium sulphate with the stoichiometric amount of carbon	1 800°C/2 hours in inert atmosphere	110
Tungsten carbide	Ammonium metatungstate with about 20 per cent carbon black	1 500°C/$\frac{1}{2}$ hour in inert atmosphere	110
	Ammonium paratungstate solution sprayed into fluidised bed of carbon black	1 200°C	102

be reduced by conducting the decomposition rapidly at high temperatures and immediately quenching the product. This is conveniently done by atomising a solution or suspension of the material to be decomposed into a flame. A wide variety of single and double oxides has been prepared by this route, as well as metallic silver and silver-palladium resulting from decomposition of the oxides (*see* Table 7).

Another method of reducing sintering is to separate the particles as they are formed. If oxalates are added to a bath of molten potassium nitrate, chemical decomposition is almost immediate and the escaping gas disperses the particles throughout the melt. The oxide particles may be recovered by pouring the melt into a large quantity of water or other solvent.[86]

The kinetics and mechanism of decomposition of many azides, oxalates, perchlorates, styphnates, permanganates, carbonates, hydrates, citrates, tartrates, acetates, hydroxides, etc. have been studied. Relatively few of these studies have been concerned with the size and nature of the products and examples of those which have are shown in Tables 7 to 11. On the other hand, a few decompositions, such as those of magnesium hydroxide, basic magnesium carbonate and basic aluminium sulphate, have been studied in some detail using electron microscopy, X-ray and electron diffraction and other techniques to examine the detailed morphological and crystallographic changes involved.[87] In addition, a comparison has been made of the grain size of magnesia produced by various decompositions, results being shown in Table 12.

Finally, it should be noted that an overall change of lattice symmetry is necessary if decomposition is to produce a fine powder. Where this condition is not met, the surface area of the product

TABLE 12

GRAIN SIZE OF MAGNESIA PRODUCED BY VARIOUS THERMAL DECOMPOSITIONS (5 HOURS HEATING)[88]

Material decomposed	Temperature	Grain size (nm)
$MgCO_3$	650°C	25 (growing)
MgC_2O_4	510°C	15 (growing)
$Mg(CH_3COO)_2$	380°C	12·5 (growing)
$Mg(OH)_2$	430°C	12 (no growth)
$4 MgCO_3 . Mg(OH)_2 4H_2O$	550°C	10 (slight growth)
$MgCO_3 . 3H_2O$	550°C	10 (slight growth)

differs little from that of the starting material, *e.g.* the dehydration of hydrous ferric oxide and of kaolinite.[116,117]

Other chemical reactions

Although other types of chemical reaction are capable of yielding a finely-divided product, there are few examples of the use of such reactions, probably because of the need to separate the products.

Fine chromium and nickel powders have been prepared[118] by the vacuum reduction of oxides with magnesium, lithium or sodium vapours. The starting oxide was of submicron size, while the metal particles were as fine as 6·8 nm in diameter. During the reduction there is a volume decrease which, as in thermal decompositions, leads to breaking and channelling of the product.

A completely different approach to the generation of a powder with a high surface area is that used for the catalytic Raney nickel. From particles of an alloy containing approximately equal weights of nickel and aluminium, the latter is extracted with caustic soda solution to leave a pyrophoric nickel powder.

CONCLUSION

A very wide range of materials has already been prepared in fine powder form, and there is no doubt that if the need arose many more substances could be prepared in this way. Both vapour phase reaction and thermal decomposition appear as versatile techniques, although the former has greater potential for the preparation of non-oxides. Quenching, which is needed to prevent particle growth in condensation, may lead to useful non-equilibrium phases being formed.

REFERENCES

1. Jenkins, I. (1961). Powder metallurgy. The manufacture and testing of metal powders, *Powders in industry,* London, Society of Chemical Industry, 246.
2. Bickle, W. H. (1961). Powder production by fine milling. *Ibid.,* 3.
3. Meloy, T. P. (1970). Fine grinding-size distribution, particle characterisation and mechanical methods. *Ultrafine grain ceramics,* edited by J. J. Burke, N. L. Reed and V. Weiss, Syracuse University Press, 17.
4. Gregg, S. J. (1968). *Chem. and Ind.,* 611.
5. Quatinetz, M., Schafer, R. J. and Smeal, C. R. (1963). The production of sub-micron metal powders by ball milling with grinding aids. *Ultrafine particles,* edited by W. E. Kuhn, New York, Wiley, 271.

6. Dachille, F. and Roy, R. (1961). Influence of displacive-shearing stresses on the kinetics of reconstructive transformations effected by pressures in the range 0–100,000 bars. *Reactivity of solids,* edited by J. H. de Boer, W. G. Burgers, E. W. Gorter, J. P. F. Huesse and G. C. A. Schuit, Amsterdam, Elsevier, 502.
7. Talbot, J. H. and Kempis, E. R. (1966). *Nature,* **197,** 66. Hofmann, U. and Rothe, A. (1968). *Zeit. anorg. Chem.,* **357,** 196.
8. Urbanski, T. (1967). *Nature,* **216,** 277.
9. Farrant, J. C. and North, R. (1957). Crushing and grinding equipment. *Chemical engineering practice Volume* 3, edited by H. W. Cremer and T. Davies, London, Butterworth, 48.
10. Strickland-Constable, R. F. (1968). *Kinetics and mechanism of crystallization,* New York, Academic Press.
11. Hirth, J. P. and Pound, G. M. (1963). Condensation and evaporation: nucleation and growth kinetics. *Prog. Mater. Sci.,* **11,** edited by B. Chalmers, Oxford, Pergamon, 1.
12. Buff, F. P. (1955). *J. chem. Phys.,* **23,** 419.
13. Cahn, J. W. (1959). *J. chem. Phys.,* **30,** 1121; Cahn, J. W. and Hilliard, J. E. (1959). *J. chem. Phys.,* **31,** 688.
14. Sutugin, A. G. and Fuchs, N. A. (1968). *J. Colloid Interf. Sci.,* **27,** (2), 216.
15. Dunning, W. J. (1955). Theory of crystal nucleation from vapour, liquid and solid systems. *Chemistry of the solid state,* edited by W. E. Garner, London, Butterworth, 159.
16. Dunning, W. J. (1969). Nucleation and growth during dehydration. *Kinetics of reactions in ionic systems,* edited by T. J. Gray and V. D. Frechette, New York, Plenum Press, 132.
17. Nielsen, A. E. (1964). *Kinetics of precipitation,* Oxford, Pergamon Press.
18. Moazed, K. L. and Pound, G. M. (1964). *Trans. Amer. Inst. Mining metall. Engrs,* **230,** 234.
19. Sinclair, D. and LaMer, V. K. (1949). *Chem. Revs,* **44,** 245.
20. Matijevic, E., Espenscheid, W. F. and Kerker, M. (1963). *J. Colloid Sci.,* **18,** 91.
21. Espenscheid, W. F., Matijevic, E. and Kerker, M. (1964). *J. phys. Chem.,* **68,** 2831.
22. Jacobsen, R. T., Kerker, M. and Matijevic, E. (1967). *J. phys. Chem.,* **71,** 514.
23. Kimoto, Kazuo and Nishida, Isao. (1967). *Japan J. appl. Phys,* **6,** 1047.
24. Hrbac, J. and Jiska, J. (1971). *Staub Reinhalt Luft,* **31,** 18; Pfender, E. and Boffa, C. U. (1970). *Rev. Sci. Inst.,* **41,** 655.
25. Greenfield, M. A., Koontz, R. L. and Hausknecht, D. F. (1971). *J. Colloid Interf. Sci.,* **35,** 102.
26. Plummer, M. (1958). *J. appl. Chem.,* **8,** 35.
27. Rooksby, H. P. (1958). *J. appl. Chem.,* **8,** 44: Das, A. R. and Fulrath, R. M. (1965). Liquid–solid transformation kinetics in Al_2O_3. *Reactivity of solids,* edited by G. M. Schwab, Amsterdam, Elsevier, 31.
28. Scheer, C. and Korman, S. (1956). The high-intensity arc in process chemistry. *Arcs in inert atmospheres and vacuum,* edited by W. E. Kuhn, New York. Wiley, 169.
29. Holmgren, J. D., Gibson, J. O. and Sheer, C. (1963). Some characteristics of arc vaporized submicron particulates. *Ultrafine particles,* edited by W. E. Kuhn, New York, Wiley, 129: (1964). *J. electrochem. Soc.,* **111,** 362.
30. Vitro Corporation, 200 Pleasant Valley Way, West Orange, New Jersey.
31. Everest, D. A., Sayce, I. G. and Selton, B. (1971). *J. Mater. Sci.,* **6,** 218.

32. Everest, D. A., Sayce, I. G. and Chilton, H. T. J. German patent 1,940,832 (30 July, 1970).
33. Barnes, W. R. and Barby, D. British patent 1,211,703 (11 November, 1970).
34. Barnes, W. R. British patent 1,211,702 (11 November, 1970).
35. Amick, J. and Turkevich, J. (1963). Electron microscopic examination of aerosols formed in a direct current arc. *Ultrafine particles,* edited by W. E. Kuhn, New York, Wiley, 146.
36. See *e.g. Vapor Deposition,* edited by C. F. Powell, J. H. Oxley and J. M. Blocher, New York, Wiley (1966). *Chemical vapor deposition of refractory metals alloys and compounds,* edited by A. C. Shaffhauser, Amer. nucl. Soc., (1967).
37. Mazdiyasni, K. S., Lynch, C. T. and Smith, J. S. (1965). *J. Amer. ceram. Soc.,* **48**, 372.
38. Lamprey, H. and Ripley, R. L. (1962). *J. electrochem. Soc.,* **109**, 713.
39. Neuenschwander, E. (1966). *J. less common Metals,* **11**, 365.
40. Barry, T. I., Bayliss, R. K. and Lay, L. A. (1968). *J. Mater. Sci.,* **3**, 229, 239.
41. Audsley, A. and Bayliss, R. K. (1969). *J. appl. Chem.,* **19**, 33.
42. Lauder, W. B. and Eichelberger, W. C. U.S. patent 3,366,443 (30 January, 1968).
    Foulds, J. T. U.S. patent 3,518,052 (30 June, 1970).
    Holden, C. B. U.S. patent 3,558,274 (26 January, 1971).
    Mas, R. J. and Michaud, A. L. U.S. patent 3,519,395. (7 July, 1970).
    Vogt, G., Wiebke, G. and Eberle, L. U.S. patent 3,560,151 (2 February, 1971).
    Jones, P. M. and Read, D. U.S. patent 3,554,708 (12 January, 1971).
    Kulling, A. and Noack, E. U.S. patent 3,539,303 (10 November, 1970).
    Dunham, W. W., Stoddard, C. K. and Rodman, H. G. U.S. patent 3,560,152 (2 February, 1971).
43. Mezey, E. J. (1966). Pigments and reinforcing agents. *Vapor deposition,* edited by C. F. Powell, J. H. Oxley and J. M. Blocher, New York, Wiley, 423.
44. Lauder, W. B. and Copson, R. L. U.S. patent 3,264,124 (2 August, 1966).
45. Butler, G. U.S. patent 3,363,981 (16 January, 1968).
46. Krinov, S. M. U.S. patent 3,363,980 (16 January, 1968).
47. Smiley, S. H. U.S. patent 3,341,320 (12 September, 1967).
48. Societa Italiana Resine. British patent 1,207,860 (7 October, 1970).
49. Thiele, K. H. and Schwartz, W. Ger. (E) Pat. 51,628 (5 November, 1966).
50. O'Connor, T. L. and McRae, W. A. U.S. patent 3,368,871 (13 February, 1968); W. E. Kuhn. (1963). The formation of silicon carbide in the electric arc. *Ultrafine Particles,* edited by W. E. Kuhn, New York, Wiley, 156.
51. Loftman, K. A. (1963). Physical characteristics and surface properties of pyrogenic oxides of silicon and aluminium. *Ibid.,* 196.
52. Case, L. E. U.S. patent 3,320,145 (16 May, 1967).
53. Wagner, E. U.S. patent 2,990,249 (27 January, 1961).
54. Kuhn, W. E. U.S. patent 3,469,941 (30 September. 1969.
55. Swaney, L. R. U.S. patent 3,485,586 (23 December, 1969).
56. Cleaver, D., Watson, W. S. and Coulthurst, A. E. British patent 1,134,782 (27 November, 1968).
57. Comyns, A. E. and Cleaver, D. British patent 1,199,811 (22 July, 1970).
58. Sayce, I. G. and Selton, B. Preparation of ultrafine refractory powders using the liquid wall furnace. *Special Ceramics* 5 edited by P. Popper (in press).
59. Everest, D. A., Sayce, I. G. and Selton, B. (1971). *Chemistry in Britain,* **7**, 318.

60. May, D. and Kolthoff, I. M. (1948). *J. phys. Chem.*, **52**, 836.
61. Kolthoff, I. M. and Bowers, R. C. (1954). *J. Amer. chem. Soc.*, **76**, 1510.
62. Brown, R. A. (1968). *Electrochem. Technol.*, **6**, (7/8), 246.
63. Gregg, S. J. and Pope, M. I. (1961). *Kolloid Zeit.*, **174**, 27.
64. Fletcher, J. M. and Hardy, C. J. (1968). *Chem. and Ind.*, 48.
65. Grimes, J. H. and Scott, K. T. B. (1968). *Powder Metall.*, **11**, (22), 213.
66. Vicarini, M. A., Nicolaon, G. A. and Teichner, S. J. (1969). *Bull. Soc. chim. (France)* 1466; (1970), 431.
67. Mazdiyasni, K. S. and Brown, L. M. (1970). *J. Amer. ceram. Soc.*, **53**, 43, 585.
    Smith, J. S., Dolloff, R. T. and Mazdiyasni, K. S. (1970). *J. Amer. ceram. Soc.*, **53**, 91.
    Brown, L. M. and Mazdiyasni, K. S. (1970). *J. Amer. ceram Soc.*, **53**, 590.
    Mazdiyasni, K. S., Dolloff, R. T. and Smith, J. S. (1969). *J. Amer. ceram. Soc.*, **52**, 523.
68. Sellars, J. W. and Toonder, F. E. (1965). Reinforcing fine particle silicas and silicates. *Reinforcement of elastomers*, edited by G. A. Kraus, New York, Interscience, 405.
69. Stöber, W., Fink, A. and Bohm, E. (1968). *J. Colloid Interf. Sci.*, **26**, 62.
70. Barksdale, J. (1966). *Titanium—its occurrence, chemistry and technology*, 2nd edition, New York, Ronald Press.
    Oster, F. U.S. patent 3,533,742 (13 October, 1970).
71. Tseung, A. C. C. and Bevan, H. L. (1970). *J. Mater. Sci.*, **5**, 604.
72. Burkin, A. R. (1967). *Metall. Revs*, **12**, 111.
    Kunda, W., Evans, D. J. I. and Mackiw, V. N. (1964). *Planseeber, Pulv metall.*, **12**, (3), 153.
73. Schuele, W. J. and Deetscreek, V. D. (1961). *J. appl. Phys.*, **32S**, 235.
74. Chalyi, V. P. and Lukachina, E. N. (1968). *Inorg. Mater.*, **4**, (2), 194.
75. Gallagher, P. K. and Schrey, F. (1963). *J. Amer. ceram. Soc.*, **46**, 567.
76. Murayama, H., Kobayashi, K., Koishi, M. and Meguro, K. (1970). *J. Colloid Interf. Sci.*, **32**, 470.
77. Murayama, H. and Meguro, K. (1970). *Bull. chem. Soc. Japan*, **43**, 2386.
78. Westinghouse Electric Corporation. British patent 1,210,900 (4 November, 1970).
79. Young, D. A. (1966). *Decomposition of Solids*, Oxford, Pergamon, 8.
80. Thomas, J. M. and Renshaw, G. D. (1967). *J. chem. Soc.*, (A), 2058.
81. Tompkins, F. C. and Young, D. A. (1956). *Trans. Faraday Soc.*, **22**, 1245.
82. Johnson, D. W. and Gallagher, P. K. (1971). *J. phys. Chem.*, **75**, 1179.
83. Dell, R. M. and Weller, S. W. (1959). *Trans. Faraday Soc.*, **55**, 2203.
84. Clayton, J. C. and Aaronson, S. (1961). *J. Chem. Engg Data*, **6**, 43.
85. Anderson, P. J. and Horlock, R. F. (1962). *Trans. Faraday Soc.*, **58**, 1993.
86. Schule, W. J. and Deetscreek, V. D. (1963). Fine particle ferrites. *Ultrafine particles*, edited by W. E. Kuhn, New York, Wiley, 218.
87. Stringer, R. K., Warble, C. E. and Williams, L. S. (1969). Phenomenological observations during solid reaction. *Kinetics of reactions in ionic systems*, edited by T. J. Gray and V. D. Frechette, New York, Plenum Press, 53, and reference therein.
88. Shin-Ichi Iwai, Hideki Morikawa, Tetsuo Watanabe and Hideki Aoki (1970). *J. Amer. ceram. Soc.*, **53**, 355.
89. Nielsen, M. L., Hamilton, P. M. and Walsh, R. J. (1963). Ultrafine metal oxides by decomposition of salts in a flame. *Ultrafine particles*, edited by W. E. Kuhn, New York, Wiley, 181.
90. Malinofsky, W. W. and Babbitt, R. W. (1961). *J. appl. Phys.*, **32S**, 237.

91. Glaister, R. M., Allen, N. A. and Hellicar, N. J. (1965). *Proc. Brit. ceram. Soc.*, **3**, 67.
92. Volmer, M. and Seydell, G. (1937). *Z. phys. Chem.*, **179**, 153.
93. Jones, K. A., Acheson, R. J., Wheeler, B. R. and Galwey, A. K. (1968). *Trans. Faraday Soc.*, **64**, (7), 1887.
94. Frost, G. B., Moon, K. A. and Tompkins, E. H. (1951). *Canad. J. Chem.*, **29**, 604.
95. Frost, G. B. and Campbell, R. A. (1953). *Canad. J. Chem.*, **31**, 107.
96. Quinn, H. W., Missen, R. W. and Frost, G. B. (1955). *Canad. J. Chem.*, **33**, 286.
97. Wheeler, R. C. and Frost, G. B. (1955). *Canad. J. Chem.*, **33**, 546.
98. Hammel, F. (1939). *Annales Chim.*, **11**, 247.
99. Ford, R. W. and Frost, G. B. (1956). *Canad. J. Chem.*, **34**, 591.
100. Richardson, J. T. and Milligan, W. O. (1956). *Phys. Rev.*, **102**, 1289.
101. Bozon-Ver'Duraz, F. and Teichner, S. J. (1968). *J. Catalysis*, **11**, 7.
102. Jordan, M. E. and Hardy, J. F. U.S. patent 3,337,327 (22 August, 1967).
103. Royen, P. and Tromel, M. (1963). *Ber. Bunsengell, phys. Chem.*, **67**, 908.
104. Dollimore, D. and Nicholson, D. (1962). *J. chem. Soc.*, 960.
105. Carruthers, J. D. and Sing, K. S. W. (1967). *Chem. and Ind.*, (45), 1919.
106. Allred, V. D., Buxton, S. R. and McBride, J. P. (1957). *J. phys. Chem.*, **61**, 117.
107. Henry, J. L. and Kelly, H. J. (1965). *J. Amer. ceram. Soc.*, **48**, 217.
108. Vincent, F. and Figlarz, M. (1969). *Les Solides finement divises*, edited by J. Ehretsmann, Paris, Documentation Francaise, 71.
109. Roginskii, S. Z., Seleznev, V. A. and Kushnerev, M. Ya. (1967). *Dokl. (Proc.) Akad. Sci. U.S.S.R. phys. Chem.*, **177**, (1), 800.
110. Cabot Corp., British Patent 1,110,386 (18 April, 1968).
111. Schroeder, H. (1967). *Zeit. phys. Chem. Leipzig*, **236**, 200.
112. Matsushita Elec. Ind. Co. (Hiromu Sasaki), U.S. patent 3,352,632 (14 November, 1967).
113. De Lau, J. G. M. (1970). *Bull. Amer. Ceram. Soc.*, **49**, 572.
114. Rettew, R. R., Wirth, D. G. and Glemzer, R. German patent 1,962,338 (12 November, 1970).
115. Adams, C. R., Voge, H. H., Morgan, C. Z. and Armstrong, W. E. (1964). *J. Catalysis*, **3**, 379.
116. Goodman, J. F. and Gregg, S. J. (1956). *J. chem. Soc.*, 3612.
117. Gregg, S. J. and Hill, K. J. (1953). *J. chem. Soc.*, 3951.
118. Arias, A. (1968). *Powder Metall.*, **11**, (22), 411.

CHAPTER 3

CHARACTERISATION OF FINE POWDERS

INTRODUCTION

The complete characterisation of a fine powder would be a task of considerable magnitude. Factors needing consideration would include:

- —chemical and phase composition and purity;
- —size, shape and roughness of primary particles;
- —surface area of primary particles;
- —number and size distribution of pores;
- —whether single or polycrystalline;
- —state of surface and internal stress;
- —concentration of defects;
- —presence of surface films and adsorbed impurities;
- —state of agglomeration.

That some attempt may now be made to determine all of these characteristics of a fine powder represents a considerable technical advance over the last twenty years or so. Prior to that time, X-ray and electron diffraction, together with measurement of surface area, probably by gas adsorption, represented the bulk of studies feasible on a powder. In the intervening years, techniques have advanced considerably. Chemical and physical methods of analysis have enabled purity to be specified with a sensitivity previously undreamed of. At the same time, there have been great advances, especially in electron microscopy. It is now possible, using this technique at suitable magnifications, to measure the size distribution of particles, to gain some information about particle shapes, to see surface roughness, to measure pore sizes and, in conjunction with selected area electron diffraction, to study the morphology of particles and, in favourable cases, to see the individual planes of atoms (Figs. 10

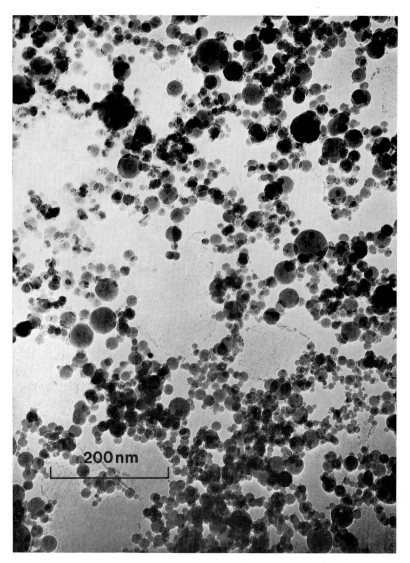

Fig. 10.   *Transmission electron micrograph of fine alumina.*

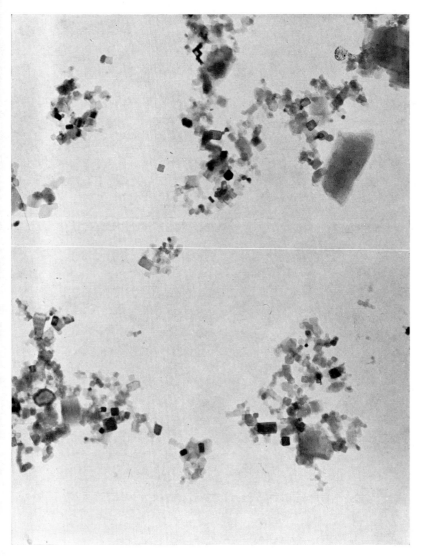

Fig. 11. *Transmission electron micrograph of fine magnesia prepared by condensation.* 4·55 cm ≡ 500 nm.

to 14). Outstanding studies have been those on carbon black[1] (the structure of which is shown diagrammatically in Fig. 15[1]) and on oxides formed by thermal decomposition.[2] These studies show what is possible with current techniques.

Some of the characterisation factors listed above will be discussed in some detail. Before this, however, a few general remarks appear to be necessary.

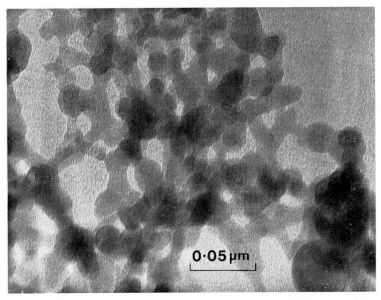

Fig. 12.   *Transmission electron micrograph of silica condensed in a moist environment.*

Processes for the preparation of a fine powder may lead to the production of metastable phases. This has already been noted above (page 7) in connection with grinding procedures and condensation from the vapour (page 12). Rapid quenching from the liquid phase by the so-called splat cooling technique has produced a number of metastable phases, as well as extended ranges of solid solution. Earlier work on metallic systems has been reviewed.[3] For oxide systems similar metastable amorphous and polymorphic phases and extended solid solution ranges have been reported.[4] It is to be expected

that condensation processes will produce a similar, but more extensive, range of metastable phases.

The surface area of a powder particle is clearly a function of size,
shape and porosity. However, without considering porosity, it is
evident that both size and shape will exert an important influence on
the activity of a particle. In cubic particles of 100 nm the number of
surface atoms is a negligible fraction of the total number of atoms,

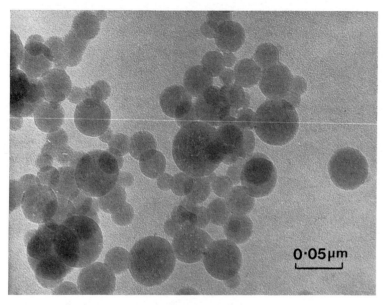

0·05μm

Fig. 13.   *Transmission electron micrograph of silica condensed in a dry
environment.*

but when the cubic particles are reduced to 10 nm, about 10 per cent
of the atoms lie in, or very close to, the surface while, in cubes of
1 nm (if it were possible to obtain them) almost 100 per cent of the
atoms would be in, or very close to, the surface.

The shape of the primary particles will determine how these
particles pack and, in many instances, will also influence chemical
and physical reactivity. Platelet particles of hexagonal close-packed
crystals of layer lattice type have many more active atoms per unit
weight than a cube-shaped particle.[5] Surface energy, which can affect

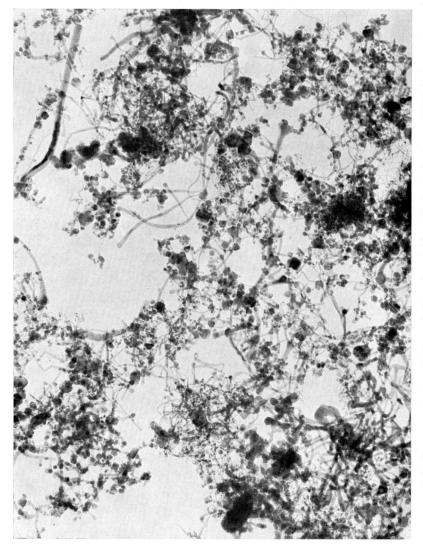

Fig. 14.   *Transmission electron micrograph of 'SiO'.* 4·27 cm ≡ 1 μm.

such properties as the sintering behaviour, is influenced by the density of atoms in the surface. In a face centred cubic structure it has been calculated that the relative surface energy of the (210) plane is 1·275 times that of the (111) plane.[5]

Particles obtained from condensation processes are often discrete and usually free from micropores, while material prepared by thermal decomposition is often highly porous and may possess comparatively

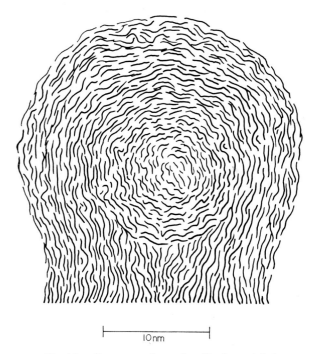

10nm

Fig. 15. *Structure of a carbon black particle.*[1]

low exterior surface. In the latter case, the particle, as seen under the microscope, consists of a number of primary particles. With a condensed powder it is often impossible to disperse the particles adequately so that selective area electron diffraction can be applied to a single particle. In these circumstances, some indication as to whether the particle is a single crystal can be obtained only by comparing the crystallite size obtained, say, by X-ray diffraction line broadening with particle size from gas adsorption or electron microscopy.

The stresses and defects in ultrafine particles are mainly defined by their history. Particles produced by condensation are believed to be largely free from stress unless the particle is made up from smaller crystallites (*e.g.* carbon black) when growth of these crystallites can cause various stresses. Powders produced by thermal decomposition, on the other hand, are stressed in the course of preparation (*see* page 25).

Virtually all fine powder particles are covered with chemically-bound or physically-adsorbed layers. Some of the consequences of this are discussed in Chapters 4 and 8.

The state of aggregation of a powder is a very complex function which is especially important in carbon blacks used for the reinforcement of elastomers and in silicas imparting thixotropy. Carbon blacks have been the subject of detailed study by electron microscopy and other techniques and methods for describing the agglomeration have been suggested and tested.[6,7]

Calculation shows that, because of the tension of the surface due to the anisotropy of bonding of surface atoms, the lattice parameter of a crystalline solid should decrease as the particles become smaller, and this has been shown to be true for magnesium oxide prepared in a vacuum.[8]

Since it is usually difficult to express the actual shape of a particle, size is often expressed as the equivalent sphere diameter calculated from the surface area assuming a non-porous particle.

PARTICLE SIZE

Particle size analysis is a subject of considerable importance, with an extensive literature[9,10,11,12,13] and, although it is true that most attention is centred on particles of micrometre size and above, much work has also been carried out on submicrometre particles.

The most comprehensive survey of the applicability and accuracy of methods for size analysis is that in progress, at the time of writing, under the auspices of the Particle Size Analysis Sub-Committee of the Society for Analytical Chemistry.[14] An initial survey of available methods and equipment has been followed by the first of a series of critical reviews—on sedimentation methods.[15]

Since many of the methods considered in these publications are

TABLE 13
METHODS OF PARTICLE SIZE ANALYSIS OF SUBMICROMETRE
POWDERS

Principle	Function measured	Approx. submicrometre size range
Centrifugal sedimentation	Equivalent diameter	>25 nm
Electron image formation		>5 nm
Conductivity change in electrolyte	Volume	>200 nm
Gas adsorption based on Harkins–Jura or BET	Surface area	
Light scattering (for approximately monodispersion)	Diameter (various)*	>10 nm
X-ray diffraction line broadening	Crystallite size	<~500 nm
Low angle X-ray scattering		

* Various average diameters depending on size range and the function measured.[16]

suitable only for larger particles, those methods useful for sub-micrometre size particles are shown in Table 13. It is important to bear in mind the function actually measured and its relationship to size.

Several other properties of materials are size-dependent and might thus (at least in principle) lend themselves to size measurement either at present or at some later stage of development. To quote just two examples, the electro-optical effect might be used for particles less than about 500 nm[17] and, for superconducting materials, measurements of magnetisation can be used to provide particle size data.[18]

Sedimentation methods

Since these methods have recently been reviewed[15] it will suffice to point out the factors involved and the limitations when applied to submicrometre particles.

Sedimentation measures the Stokes diameter, *i.e.* the diameter of a sphere with the same density and same free falling velocity in the fluid used as the particles under study. With submicrometre particles natural settling becomes so slow that the assumption of undisturbed separate streamlined flow for each particle is no longer true. The terminal velocity of the particles becomes comparable with that of the inevitable convection currents due to thermal gradients or liquid displacement.[19] Centrifugal settling is used to overcome these difficulties. The Stokes equation then becomes:

$$d^2 = \frac{18\eta \ln x/x_0}{(\rho - \rho_L)w^2 t}$$

where:

d = particle diameter
η = fluid viscosity
ρ = particle density
ρ_L = fluid density
t = time
w = angular velocity of rotation
x = distance of particle diameter, d, from axis of rotation
x_0 = radius of rotation of this particle at zero time.

The Stokes equation is valid only for laminar flow of individual particles so that the particles must be completely dispersed and there must effectively be no convection currents in the fluid during sedimentation.

When the suspension is applied carefully to the top of the fluid column all the particles have the same initial radius of rotation and the Stokes equation can be used directly (two layer technique). In this technique 'streamers' tend to form. These are aggregates of powder particles formed when the suspension breaks through into the column fluid and the Stokes equation is no longer obeyed. The formation of streamers may be reduced by placing a thin layer of miscible liquid of density intermediate between that of the suspension and the column fluid on top of the column before adding the suspension.[20] This is the so-called 'buffered line start' technique. When the suspended powder is initially dispersed homogeneously throughout the fluid, a differential equation describes the distribution. This equation can be developed for variable x,[21] or variable t.[22,23] The latter is more readily realised experimentally, but does not permit of an exact solution. Numerical solutions have been developed. Details of commercially-available apparatus for sedimentation analysis have been given.[13]

To enable still finer particles to be sedimented, the ultracentrifuge has been used. To determine the amount of sedimented material, optical methods based on schlieren optics or interferometry or absorption are usually employed.[24,25] In this way, the size range from 25 nm to 10 μm can be covered and Stokes' law applied.

Electron image formation

The electron microscope provides the only direct method of measuring particle size. All other methods make use of the variation in some property which itself varies with particle size. In addition, some assessment can be made of particle shape (*see* below).

Operation of the electron microscope is a highly-developed art with its own literature (*see*, for example, reference 26). The need to separate the images of particles sufficiently well to measure each individually, together with the limited penetration of the electron beam, means that a high degree of dispersion of the powder is required. During dispersion the existing particles must not be fractured, so operations, such as grinding to break up aggregates, are to be avoided. The usual method of dispersion involves ultrasonic vibration of a suspension in a low viscosity solvent which is often volatile. There is usually an optimum period for which vibration should be continued, after which time reaggregation becomes predominant. A small drop of the dispersion is then transferred to a carbon grid on the microscope stage, either by an atomising spray or with the aid of a micropipette. Alternatively, the dispersion may be made in a solution of cellulose acetate in acetone and a drop of the solution allowed to spread on a glass slide. The film formed on evaporation may then be cut into sections, floated off the slide with water and transferred directly to the microscope grid.[27]

Powders prepared by condensation often consist of discrete particles and can then be dispersed adequately. Thermal decomposition, on the other hand, usually leads to the formation of a network type of material where the individual crystallites are in very limited contact with their neighbours.

When the decomposition temperature is such that some sintering occurs these contacts may become so strong that a useful dispersion cannot be obtained except by grinding to cause fracture.

The choice of fields of view on the microscope may be either by random or by operator selection.[27] The number of particles to be measured and fields to be scanned are governed by statistical considerations.[28] Automated methods for scanning and sizing electron micrographs are available,[29] but may be applied only when the particles do not overlap and are not in close contact.

When deriving particle size or any other information from electron micrographs it must, at all times, be remembered that an exceedingly small amount of material is under examination, and that this sample may not be typical. Thus, if in the preparation of the dispersion by ultrasonic vibration there is an appreciable sediment not in suspension, material which is in suspension is obviously not typical of the whole. Settled material may merely be agglomerated in some way, or may consist of larger primary particles of different shapes or phases.

Change in the electrical conductivity of an electrolyte

A dispersion of the powder in a suitable electrolyte is passed through a narrow orifice. In so doing, each particle causes a change in the electrical resistance across the orifice by displacing an equivalent volume of electrolyte. In the commercial instrument (Coulter Counter) the resistance change is converted to a voltage pulse of amplitude proportional to the particle volume. By using various threshold voltages in the counting, and, if necessary, a series of different orifices, a size distribution based on particle volume may be built up. Calibration is made using a standard mono disperse powder, but different standards yield different instrument constants. There is a tendency to undercount fine particles and a high degree of dispersion is required. The commercial instrument is suitable for use with particles greater than about 0·8 μm in diameter. A very detailed critical review has recently been made of the operation and limitations of the Coulter Counter, and an exhaustive list of published work on the use of the instrument is also available.[30,31] In spite of its limitations, this method is useful since the reproducibility of a given instrument is very good over a long period of time and meaningful size comparisons may be made.[32]

Light scattering

As a technique for the determination of the particle size of fine powders, light scattering has been somewhat neglected. However, this may soon be rectified following the introduction into the technique of the laser beam.[33]

The general Mie theory of scattering shows a dependence of the more readily-measured parameters such as angular variation, polarisation ratio, high order Tyndall spectrum and turbidity on particle diameter (d), angle of scatter (θ) and m, the ratio:[34]

$$\frac{\text{refractive index of particle}}{\text{refractive index of medium}}$$

There is also a lesser dependence on the concentration of the suspension.[35] The more simple theories of light scattering are applicable only over restricted values of d and m.[36] Recent use of computers has enabled a series of calculations to be made connecting the parameters concerned in Mie scattering.[37-40] Dispersions of essentially mono-sized latex spheres have been used to correlate the particle size obtained from these calculated relations with that obtained by electron microscopy.

When the particle size is considerably less than the wavelength of light (Rayleigh region) the angular dependence of scatter is a monotonic function. As particle size increases, this changes to a wave function with a series of alternate maxima and minima. The angular position of a given order of maximum or minimum is related to particle size by a function of the type:

$$d \sin \theta/2 = k_\lambda$$

where λ is the wavelength of light in the medium and k is dependent on m. Whilst this function is probably applicable when d exceeds 0·05 μm, experimental study has been restricted to $d > 0·2$ μm. [39,41,42] With the aid of this technique, a narrow size distribution may also be explored.[43]

Tyndall spectra are produced when particles of size comparable with the wavelength of light, in the form of a dilute suspension, are irradiated with white light. Bands of colour (red or green) are then visible at certain angles. If the relative intensities of red and green scattered light (extrapolated to zero dispersion concentration) are plotted against θ the curve is similar in shape to that for the angular dependence of scattering with one or more maxima and minima. The relation:

$$d \sin \theta/2 = k$$

may be used to calculate d.[44]

The polarisation ratio at 90° is the ratio of the intensities of the horizontal and vertical components of the scattered beam, and can be used in a manner similar to that described above.[45]

Turbidity measurements on a mono disperse suspension have long been used to investigate particle size. There are several precautions to be taken concerning scattered stray light, 'corona' at the exit of the cell and the solid angle of scattered light reaching the photocell, etc., but these have often been ignored.[46]

Application of general scattering theory to coated spheres,[47] cylindrical particles[48] and to aggregates of smaller particles[49] have been reported.

Since all these applications of light scattering yield some sort of average value for particle size, use is normally restricted to fairly narrow size distributions. Attempts have been made, apparently with some success, to extend the use of these techniques to give a particle size distribution[38,50] for a fairly narrow size range, but this technique does not seem to have been widely applied.

Apparatus which can be used for these measurements has been discussed in many of the papers referred to above, as well as in standard texts.[10,12]

X-ray diffraction broadening

The broadening of X-ray diffraction lines is primarily a measure of the departure from single crystal perfection and regularity in the specimen. Thus, as well as small size, lattice strain, dislocations, impurities and other defects lead to line broadening. Nevertheless, as the only method which gives the size of the primary crystallites, irrespective of how they are aggregated or sintered, X-ray diffraction broadening is of great value with the finest of powders. The technique has been reviewed recently.[51,52]

The classical Scherrer relationships are used to describe the effect of average particle size and of strain on the width of an X-ray diffraction line:[53]

$$\beta_p = K\lambda/\varepsilon \cos \theta$$
$$\beta_s = \eta \tan \theta$$

where:

β_p = broadening due to particle size
β_s = broadening due to strain
ε = a measure of particle size
K = a constant (shape factor)
η = the strain function
λ = the wavelength
θ = the Bragg angle.

From the experimental line width, usually measured on a diffracto-meter, it is necessary to remove the instrumental broadening in order to obtain the true broadening β for use in the Scherrer relations. This can be done either by somewhat empirical methods[53] or by using Fourier transform theory (e.g. reference 54). For broadening due to particle size $\beta_p \propto \lambda/\cos \theta$; for that due to strain $\beta_s \propto \tan \theta$.[55] These relations make possible the determination of the cause of broadening when this results from a single source, but no satisfactory method has yet been developed for assigning the proportions of mixed broadening. Having found a value for β_p and selected an appropriate value for K $(0\cdot95-1\cdot15)$[53] the value of ε can be calculated. Although this procedure is satisfactory when a comparison of particle sizes of the same materials is required, it does not give accurate absolute values. For example, some workers have found that β is a function of tan θ for any one reflection (change of λ), irrespective of the nature of the broadening.[56]

More recently, the pure profile of an individual line has been represented by a Fourier series. From the Fourier coefficients the broadening effects of size, strain, etc., can be separated.[57]

A comparison of various ways of measuring the average particle size from broadening of the (220) reflection for MgO has shown generally good agreement.[58]

Particle size distribution can also be obtained from X-ray diffraction broadening, at least for materials with cubic unit cells. By using (0 0 l) reflections, the intensity of the reflected line can be calculated in terms of the height of columns of unit cells perpendicular to the (0 0 l) planes which represent the crystallites.[57,59] A comparison of the size distribution of precipitated cobalt ferrite obtained in this way showed satisfactory agreement with that obtained from electron microscopy.[60]

Although these broadening methods may be used on particles of up to 500 nm diameter, it is only in the smaller particles of diameter less than about 20 nm that the size effect predominates over other causes of broadening.

Low angle X-ray scattering may also be used for determining the surface area of particles below about 100 nm diameter.[61] Such scattering arises from all grains, irrespective of their orientation, and is much more intense than the broadening of a Debye–Scherrer diffraction line. Several procedures are available for calculating surface area from the scattered intensity data. When results have been compared with other techniques, agreement is usually satisfactory, or can be explained by the nature of the sample.[61]

Gas adsorption

Gas adsorption measures surface area, from which particle size can be obtained only as an average value when some assumption is made concerning particle shape and the presence or absence of pores. Fortunately, adsorption methods themselves can provide some information on pores and fissures. Because the surfaces of fine powders are often of considerable importance, surface areas are valuable in their own right.

All general books (*e.g.* references 10, 11 and 12) on particle size analysis contain chapters on the theory and/or practice of gas adsorption, and there are whole books devoted exclusively to this subject.[62,63] For this reason, a general consideration only will be given here. Particle shape and porosity are discussed later.

The determination of surface area by gas adsorption depends on finding the amount of gas which represents monolayer coverage of the sample surface and assuming a value for the cross-section of the gas molecule. The intensity of the interaction between a solid surface and adsorbed gas will change when monolayer coverage is complete, but since coverage does not occur with complete uniformity (solid surfaces are energetically heterogeneous), the change is not often sharp.

The first quantitative theory considered monolayer adsorption only, but this was replaced by the well-known BET isotherm which takes account of multilayer adsorption.[64] The BET isotherm has been criticised on a number of grounds,[65,66] but, in spite of proposed improvements, has not been completely replaced.

Many thousands of isotherms showing the relationship between the relative pressure (p/p_0 where p_0 is the saturated vapour pressure of the adsorbate at the temperature of operation) and the amount of gas adsorbed have been plotted over the years. The vast majority of these fall into one or other of the five shapes considered by BET. These are shown in Fig. 16. The BET isotherm may be expressed as:

$$\frac{p/p_0}{V_{ads}[1 - (p/p_0)]} = \frac{1}{V_m C} + \frac{C - 1}{V_m C} p/p_0$$

where V_{ads} is the NTP volume of gas adsorbed at p/p_0, V_m is the monolayer coverage volume and C is a constant. When $2 > C > 1$, the isotherm is type III; as C increases above 2, isotherms of type II are formed. Types IV and V are similar, at lower p/p_0, to types II and III, respectively, but, at higher values of p/p_0, usually show hysteresis and as p/p_0 approaches unity, different shapes emerge. These differences at higher values of p/p_0 are considered to be due to the presence of pores in isotherms of types IV and V.

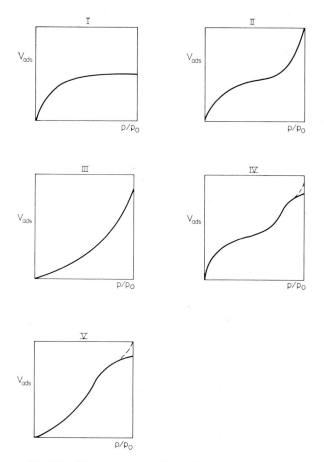

Fig. 16.   *The main types of gas adsorption isotherm.*[62]

The sharpness of the 'knee' at fairly low p/p_0 for type II and IV isotherms is a measure of the change in the net heat of adsorption on completion of the monolayer. This does not mean that the knee or the point of inflexion necessarily occur at the point of monolayer coverage. Thus, if the knee is very sharp and occurs at very low p/p_0, some localised adsorption, such as chemisorption, may be suspected. Under such conditions the surface area calculated from gas adsorption is open to doubt. Where the point of inflexion is readily located without the knee being exceptionally sharp, fair confidence may be placed in surface area values from isotherms of type II. When it has been possible to check the surface area by an independent measurement, for example, on a thin metal film where direct measurement is possible, agreement is good if an entirely reasonable value of about 1·5 is assumed for the roughness coefficient.[67,68] To calculate surface area from particle size obtained from electron microscopy, a shape function must be assumed. When this function is known with fair assurance, agreement with the surface area from gas adsorption is as good as can be expected for type II isotherms.

When the isotherm is of type III great care is necessary in accepting values for the surface area. Two factors follow from the shape of the isotherm. Firstly, adsorbate–surface interaction appears to be less important than adsorbate–adsorbate interaction (the more molecules adsorbed, the easier it is for further molecules to be adsorbed). Secondly, at low p/p_0 where the application of the BET isotherm may be expected to be more valid, the volume of gas adsorbed is very small and difficult to measure accurately.

In isotherms of types IV and V the effects of the existence of pores (hysteresis and change of shape) are seen only at p/p_0 larger than that corresponding to monolayer formation. For surface area determination these isotherms are subject to the same restrictions as types II and III, respectively.

Considerable difficulty is encountered in interpreting isotherms of type I, which often accompany chemisorption. Sometimes the original Langmuir isotherm[69] is used, but this has led, in one instance, to the conclusion that over 90 per cent of the atoms in a sample of charcoal were on the surface.[70] Probably the least unreliable surface area to be obtained from a type I isotherm is that calculated from the so-called 'Henry's Law' region of very low p/p_0, but even then there is considerable uncertainty and inaccuracy.

All these remarks about the relevant reliability of surface area measurements relate to absolute values. Where the aim is a comparison of a series of the same type of isotherm, greater reliance may be placed on the results.

Experimentally, the data for the isotherm may be obtained from either the classical static system—using volumetric or gravimetric measurement of gas quantities—or from a dynamic system which is related to gas chromatography. Suitable apparatus and a discussion of the relative merits of these systems have been given in the standard works mentioned above (see also references 71 and 72).

The repeatability of the actual measurements may be good. For example, a coefficient of variation of about 1 per cent has been reported for areas greater than 1 m^2/g.[73] Where considerable differences occur, these may be due to differences in the outgassing procedures.

Whilst nitrogen is the most widely-employed adsorbate, krypton is much used for smaller surface areas because of the lower value of p_0. However, uncertainty as to the correct value for p_0 places doubt on the absolute values of areas obtained with the latter adsorbate. At liquid nitrogen temperature, krypton is well below its triple point. The logical use of the saturated vapour pressure of the solid as p_0, leads to unusually sharp upturn on the isotherm as p/p_0 approaches unity. Consequently, the saturated vapour pressure of the supercooled liquid is usually taken for p_0. As well as the practical objection that this leads to many plots which, when converted to the usual form (see below), are not linear,[74] the validity of the use of this value for p_0 has been questioned on theoretical grounds.[75] On the other hand, many workers report good agreement between the area of a solid using nitrogen and that using krypton and the accepted cross-sectional areas of the adsorbed molecules.[76,77]

According to the BET isotherm, experimental adsorption data, when expressed as a plot of p/p_0 against $(p/p_0)/\{V_{ads}[1 - (p/p_0)]\}$, should yield a straight line. From such a plot the monolayer volume $V_m = [1/(\text{slope} + \text{intercept on the ordinate})]$. In order to use a linear portion of the plot, it may be necessary to move somewhat outside the usual limits $0.25 > p/p_0 > 0.05$. To calculate surface area from V_m, a value must be assumed for the cross-section of the adsorbate molecule. For nitrogen 0.16 nm^2 is usually used and recommended values are available for many other adsorbates.[77]

Heat of immersion

If a comparative value of particle surface area only is required, a measurement of the heat of immersion may be suitable. The heat effect is small so a sensitive calorimeter is required. If an absolute measure of surface area is required the difficulties are greater. The liquid must then be chosen so that chemical reaction is minimal and a standard value must be available for the particular solid–liquid pair in use so that conversion from heat to surface area may be made. The presence of strain in the sample can affect the heat of immersion.

Where there may be a chemical reaction, for example, $Al_2O_3-H_2O$, a wide range of adsorption energies is found.[78] This type of behaviour has been interpreted in terms of chemisorption of surface hydroxyl groups and/or hydrogen bonding.[79]

Adsorption from solution

The adsorption of a dyestuff from a suitable solution is often used to ascertain surface area because the amount of dyestuff adsorbed can easily be determined. When a comparison is to be made between a series of samples of the same composition, or when a standard sample of the same composition as the unknown is available, this procedure may be valuable. In other circumstances, great care is needed in the selection of solvent and solute so that separate adsorption isotherms can be calculated from the combined data, *i.e.* there must be strong preferential adsorption of dyestuff or other solute. If large dyestuff molecules are used as solute there is difficulty in deciding on a value for the cross-sectional area, which varies widely with the molecular orientation of the dyestuff. If the solute[10,80] contains polar groups, adsorption on a polar adsorbent will be enhanced.

EXTERIOR SHAPE

In the present context, particle shape may be divided into exterior shape and the size distribution of internal pores, if present.

The exterior shape of fine powder particles usually reflects the method or conditions of preparation. Thus it has been pointed out above that material prepared by thermal decomposition shows the relic of the crystal shape of the starting material with each macro crystal being itself porous and composed of many ultimate crystallites. Particles resulting from condensation are often spherical but can sometimes be angular. The distinction has been ascribed to whether condensation occurred via the liquid or directly to the solid,[81] but other causes cannot be ruled out. Water-soluble particles can change shape on exposure to atmospheric humidity, supposedly due to partial solution in adsorbed moisture followed by recrystallisation. Sodium chloride, which condenses as spheres, changes to cube-shaped particles on exposure to air for a few hours.[82]

Determination of particle shape

Electron microscope examination of a suitable dispersion of a fine powder will give information on particle shape or cross-section perpendicular to the viewing direction. Standard shadowing procedures may be useful in obtaining information on shape in the third dimension. Should the particles be plate-shaped it may not be possible to see any stacked on end on electron micrographs, either because the method of preparation causes the plates to lie flat, or because the plates are too thin to be seen.[83]

Where the dispersion and particle size are both suitable it may be possible to determine, by electron diffraction, the crystal orientation of several particles. If these prove to be similar it suggests that plate-like crystals have been laid down on the microscope grid.

Scanning electron microscopy can yield direct and valuable information on the shape of large particles,[84] but the limited resolution now available, ($\sim$25 nm), coupled with the fact that it is sometimes necessary to deposit a conducting coating over the particle to avoid charging effects, means that for the finer powders this technique is of little value.

The use of X-ray line broadening for the determination of particle size has been discussed above. In the event that some diffraction lines are markedly more broad than others it may be found for example that lines of the type (0 0 l) are very broad while lines where l is zero are sharp. Particle dimensions calculated from these two types of diffracted beam will then be very different and will show a plate-like shape. Lath shapes can be identified in the same way.[53]

Pore sizes

It was mentioned above that the presence of pores affects the shape of isotherms for gas adsorption, changing types II and III to types IV and V, respectively. It was also mentioned that hysteresis is typical of the presence of pores.

Methods for calculating the distribution of pore sizes are based on the Kelvin equation:

$$\ln p/p_0 = -\frac{2V\gamma}{r\mathrm{RT}} \cos \varphi$$

where V is the molar volume of liquid adsorbate filling the pores, γ is its surface tension, r the pore radius and φ the angle of contact between liquid and pore walls (this cannot be measured so it is usually assumed that the liquid wets the pore walls and cos $\varphi = 1$). The Kelvin equation was originally developed to account for the variation of vapour pressure over surfaces of varying curvature and cannot be expected to give the highest accuracy when applied to the filling of pores.

The most simple procedure for calculating the pore size distribution assumes that all the gas adsorbed has gone into pores. For each value of p/p_0, a value of r is obtained from the Kelvin equation and a value of V_{ads} from the adsorption isotherm. Then V_{ads} is the volume of all pores filled and of radius up to r. The slope of a plot of V_{ads} against r gives the pore size distribution. This procedure does not take into account the increase in the thickness of the adsorbed layer on the walls of pores of radius greater than r when p/p_0 is increased by an increment. This last factor is complex since the area exposed changes as pores fill or empty. Fairly simple approximations have been proposed to deal with this situation.[12,62] These make use of the so-called t plot,[85,86] the methods of plotting and usefulness of which have recently been reviewed.[87]

Varying with p/p_0, t is the thickness of the adsorbed layer on pore walls. It is found for a given value of p/p_0 as the product of the thickness of a molecular layer of liquid (0·35 nm for nitrogen) and the ratio of the volume adsorbed on the porous solid at p/p_0 to the monolayer volume on the non-porous solid. The latter can either be evaluated experimentally or read off (for nitrogen as adsorbate) from a 'master' isotherm. While surface is freely available and multi-layers can form on all surfaces, a plot of t against p/p_0 is linear. If the slope becomes greater than the linear value, capillary condensation is occurring; if it becomes less, the available surface is decreasing owing to the filling of pores. To put this information on to a quantitative basis some assumption must be made concerning pore shape. The usual assumption is that pores are either slit-shaped or cylindrical. To take the former as an example,

$$d = r_{\mathrm{K}} + 2t$$

where d is the distance apart of the sides and r_{K} is the Kelvin radius. At $p/p_0 = 1$ all pores, and some intergranular space, is filled with liquid adsorbate. Taking the desorption isotherm and dividing it into equal steps in p/p_0, ΔV_{ads} can be read off for each step and t calculated for the mean value of p/p_0 of each step and r_{K} calculated from the Kelvin relation. Using the assumption on pore shape, d

can be found for each step and then $\Delta V_{\text{ads}} = \frac{1}{2}d\,\Delta S$ where ΔS is the change in available surface across the step in p/p_0. Values of ΔV and ΔS, both cumulative, can be used to interpret pore sizes.[88]

While t plots (examples of which are shown in Fig. 17) are valuable in studying macroporosity, their use to characterise microporosity must be treated with caution due to marked sensitivity to small changes in the 'standard' isotherm.[89]

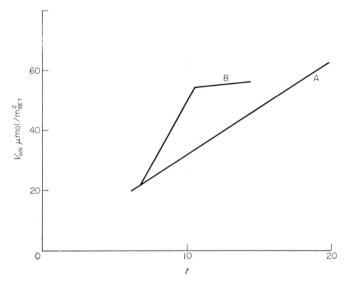

Fig. 17.   t plots for alumina. A: non-porous, B: porous: initial filling of pores with subsequent reduction in available surface.

Attempts have been made to give a rough description of pore shapes from the shape of the hysteresis loop.[90] Three types of major interest have been identified and are shown in Fig. 18.

In a similar vein attempts have been made to correlate changes in adsorption isotherms with capillary condensation of a known amount of water in saddles between fine powder spheres of equal sizes.[91] In this way, values up to 12 have been found for the co-ordination number of particles in pressed pellets.[92] However, co-ordination numbers vary according to whether the calculation is based on porosity, free surface or water film surface, and this can be accounted for if the particle surfaces are rough.[91]

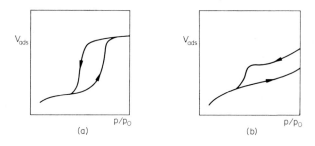

(a)   (b)

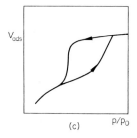

(c)

Fig. 18.   *Gas adsorption hysteresis loops for various pore shapes.*[62] *(a)*
Open cylindrical pores of fairly uniform size. (b) Slit-shaped pores between
parallel plates. (c) Ink-bottle or tubular pores with a constriction.

A comparison has been made between the pore sizes of silica gel,
as obtained by low-temperature adsorption of nitrogen, and that
given by electron microscope examination of thin edges of the gel.[93]
The agreement was fair, but more detail on pore shape was obtained
from the electron micrographs which, of necessity, showed only pores
on thin edges.

A more direct method of measuring pore sizes is by using the
mercury porosimeter in which pores are filled directly by the application of elevated pressures. The pressure required to force a liquid
into a pore of radius r is given by the Young–Laplace equation:

$$\Delta P = \frac{2\gamma}{r} \cos \varphi$$

For mercury, pores of diameters greater than 20 nm will be filled
at about 700 atmospheres pressure while about 3 500 atmospheres
pressure will be required to fill pores of 4 nm diameter. Since the

relationship of surface tension γ to r is uncertain at small values of r, pores of diameter less than 4 nm are not usually studied.

Experimentally it is necessary only to measure the pressure and volume of mercury forced into the sample. Pores are usually assumed to be cylindrical. Mercury porosimeters are available commercially; the principles have been described[94] and applications to the field of materials reviewed.[95]

REFERENCES

1. Hess, W. M., Burgess, K. A., Lyon, F. and Chirico, V. (1968). *Kaut. u. Gummi Kunstst.*, **21**, 689.
   Hess, W. M., Ban, L. L. and McDonald, G. C. (1969). *Rubber Chem. Technol.*, **42**, 1209.
   Harling, D. F. and Heckman, F. A. (1969). *Mater. Plast. Elastomeri*, **35**, (1), 80.
2. Moodie, A. F. and Warble, C. E. (1967). *Phil. Mag.*, **16**, 891.
   Stringer, R. K., Warble, C. E. and Williams, L. S. (1969). Phenomenological observations during solid reactions. *Kinetics of reactions in ionic systems*, edited by T. J. Gray and V. D. Frechette, New York, Plenum Press, 53.
3. Giessen, B. C. (1966). Rapidly quenched (splat-cooled) alloys. *Strengthening mechanisms*, edited by J. J. Burke, N. L. Reed and V. Weiss, Syracuse University Press, 273.
4. Serjeant, P. J. and Roy, R. (1967). *J. Amer. ceram. Soc.*, **50**, 500; *J. appl. Phys.*, **38**, 4540.
5. Kuhn, W. E. (1969). The role of morphology and activity in the consolidation of ultra fine particles. *Les solides finement divises* edited by J. Ehretsmann, Paris, Documentation Francaise, 91.
6. Medalia, I. A. (1969). *J. Colloid Interf. Sci.*, **24**, 393; (1970) **32**, 115; (1971) *Powder Technology*, **4**, 117.
7. Medalia, I. A. and Heckman, F. A. (1971) *J. Colloid Interf. Sci.*, **36**, 173; (1969) *Carbon*, **7**, 567.
8. Guilliat, I. F. and Brett, N. H. (1969). *Trans. Faraday Soc.*, **65**, 3328.
9. Allen, T. (1970). *Particle size analysis bibliography No.* 1 1968–70. Newcastle upon Tyne, Powder Advisory Centre.
10. Allen, T. (1968). *Particle size measurement*, Chapman Hall, London.
11. Irani, R. R. and Callis, C. F. (1963). *Particle size: Measurement, interpretation and application*, Wiley, New York.
12. Orr, C. and Dallavalle, J. M. (1959). *Fine particle measurement*, Macmillan, New York.
13. Proceedings of a conference on Particle Size Analysis, Bradford, September 9–11, 1970. To be published by the Society for Analytical Chemistry.
14. *Analyst* (1963) **88**, 156.
15. *A critical review of sedimentation methods* (1968), Society for Analytical Chemistry, London.
16. Meehan, E. J. and Beattie, W. H. (1960). *J. phys. Chem.*, **64**, 1006.
17. Kas, H. H. and Bruckner, R. (1970). *Z. angew. Phys.*, **29**, 64.
18. Lewis, R. T. (1968), *J. Colloid Interf. Sci.*, **26**, 361.

19. Kaye, B. H. and Jackson, M. R. (1970). Problems of characterising fine powders. *Ultrafine grain ceramics,* edited by J. J. Burke, N. L. Reed and V. Weiss, Syracuse University Press, 63.
20. Beresford, J. (1967). *J. Oil Colour Chemists Assoc.,* **50,** 594.
21. Bradley, D. (1962). *Chem. Process Engg.,* **43,** 591, 634.
22. Kamack, H. J. (1951). *Analyt. Chem.,* **23,** 844.
23. Slater, C. and Cohen, L. (1962). *J. sci. Instrum.,* **39,** 614.
24. McCormick, H. W. (1964). *J. Colloid Sci.,* **19,** 173.
25. Aeijelts Averink, J. W., Reerink, H., Boerma, J. and Jaspers, W. J. M. (1966). *J. Colloid Interf. Sci.,* **21,** 66.
26. *Techniques for electron microscopy,* edited by D. H. Kay. Second edition, Blackwell, Oxford, 1965.
27. Crowl, V. T. (1967). Particle size analysis by counting from electron micrographs. *Particle size analysis,* London, The Society for Analytical Chemistry, 36.
28. Herdan, G. (1953). *Small particle statistics,* Elsevier, Amsterdam.
29. Fisher, C. (1967). The Metals Research image analysing computer. *Particle size analysis,* The Society for Analytical Chemistry, London, 77.
30. Allen, T. and Marshall, K. *A critical review of the Coulter Counter.* To be published by the Society for Analytical Chemistry, London.
31. Coulter Electronics Ltd. *Coulter Counter Industrial Bibliography.* Dunstable, Beds., 1970.
32. Allen, T. (1967). Particle size measurements and their significance. *Pigments: an introduction to their physical chemistry,* edited by D. Patterson, Elsevier, Amsterdam, 102.
33. Heyder, J., Roth, C. and Stahlhofen, W. Particle size analysis of airborne particles smaller than 0·3 μm in diameter, reference 13, paper 5.
34. Van De Hulst, H. C. (1957). *Light scattering by small particles,* Wiley, New York.
35. Napper, D. H. and Ottewell, R. H. (1964). *J. Colloid Sci.,* **19,** 72.
36. Heller, W. and Pangonis, W. J. (1957). *J. chem. Phys.,* **26,** 498.
37. Pangonis, W. J., Heller, W. and Economou, N. A. (1961). *J. chem. Phys.,* **34,** 960, 971.
38. Kerker, M., Matijevic, E., Espenscheid, W. F., Farone, W. A. and Kitani, S. (1964). *J. Colloid Sci.,* **19,** 213.
39. Kerker, M., Farone, W. A., Smith, L. B. and Matijevic, E. (1964). *J. Colloid Sci.,* **19,** 193.
40. Kerker, M. and Matijevic, E. (1961). *J. opt. Soc. Amer.,* **51,** 87.
41. Maron, S. H. and Elder, M. E. (1963). *J. Colloid Sci.,* **18,** 107.
42. Maron, S. H., Pierce, P. E. and Elder, M. E. *Ibid.,* 391.
43. Clark, R. J. (1969). *Appl. Polym. Symposium,* **8,** 207.
44. Maron, S. H. and Elder, M. E. (1963). *J. Colloid Sci.,* 18, 199.
45. Maron, S. H., Elder, M. E. and Pierce, P. E. *Ibid.,* 733.
46. Heller, W. and Tabibian, R. M. (1957). *J. Colloid Sci.,* **12,** 25.
47. Kerker, M. and Matijevic, E. (1961). *J. opt. Soc. Amer.,* **51,** 506.
48. Kerker, M., Farone, W. A. and Espenscheid, W. F. (1966). *J. Colloid Sci.,* **21,** 459.
49. Hiemenz, P. C. and Vold, R. D. *Ibid.,* 479.
50. Kerker, M., Darby, E., Cohen, G. L., Kratohvil, J. and Matijevic, E. (1963). *J. phys. Chem.,* **67,** 2105.
51. Hillard, J. E., Cohen, J. B. and Paulson, W. M. (1970). Optimum Procedures for determining ultra fine grain sizes. *Ultrafine grain ceramics,* edited by J. J. Burke, N. L. Reed and V. Weiss, Syracuse University Press, 73.

52. Oel, H. J. (1969). Crystal growth in ceramic powders. *Kinetics of reactions in ionic systems,* edited by T. J. Gray and V. D. Frechette, Plenum Press, New York, 249.
53. Klug, H. P. and Alexander, L. P. (1954). *X-ray diffraction procedures,* Wiley New York, 491.
54. Stokes, A. R. (1948). *Proc. phys. Soc.,* **61,** 382.
55. Hall, W. H. (1949). *Proc. phys. Soc.,* **62A,** 741.
56. Williamson, G. K. and Hall, W. H. (1953). *Acta metall.,* **1,** 22.
57. Warren, B. E. and Averbach, B. L. (1950). *J. appl. Phys.,* **21,** 595.
58. Guilliatt, I. F. and Brett, N. H. (1970). *Phil. Mag.,* **21,** 671; (1969). *J. Brit. ceram. Soc.,* **6,** 56.
59. Smith, V. H. and Simpson, P. G. (1965). *J. appl. Phys.,* **36,** 3285.
60. Hasegawa, K. and Sato, T. (1967). *J. appl. Phys.,* **38,** 4707.
61. Eanes, E. D. and Posner, A. S. (1967). Small-angle X-ray scattering measurements of surface areas. *The solid-gas interface,* edited by E. A. Flood, Edward Arnold, London, **2,** 975.
62. Gregg, S. J. and Sing, K. S. W. (1967). *Adsorption surface area and porosity,* Academic Press, London.
63. Young, D. M. and Crowell, A. D. (1962). *Physical adsorption of gases,* Butterworth, London.
64. Brunauer, S., Emmett, P. H. and Teller, E. (1938). *J. Amer. chem. Soc.,* **60,** 309.
65. Hüttig, G. F. (1948). *Mh. Chem.,* **78,** 177.
66. Harkins, W. D. and Jura, G. (1943). *J. chem. Phys.,* **11,** 430.
67. Rhodin, T. N. (1950). *J. Amer. chem. Soc.,* **72,** 5691; (1953). *J. phys. Chem.,* **57,** 1437.
68. Bowers, R. (1953). *Phil. Mag.,* **44,** 467.
69. Langmuir, I. (1916). *J. Amer. chem. Soc.,* **38,** 2221.
70. Culver, R. V. and Heath, N. S. (1955). *Trans. Faraday Soc.,* **51,** 1569.
71. BS 4359, Part 1, 1969.
72. Robens, E., Sandstede, G. and Walter, G. (1969). *Vide,* **24,** 266.
73. Crowl, V. T. (1967). Surface area measurement by low temperature nitrogen adsorption. *Particle size analysis,* Society for Analytical Chemistry, London, 288.
74. Malden, P. J. and Marsh, J. D. F. (1959). *J. phys. Chem.,* **63,** 1309.
75. Singleton, J. H. and Halsey, G. D. (1955). *Canad. J. Chem.,* **33,** 184.
76. Medema, J. and Houtman, J. P. W. (1969). *Anal. Chem.,* **41,** 209.
77. McClellan, A. L. and Hornsburger, H. F. (1967). *J. Coll. Interface Sci.,* **23,** 577.
78. Venable, R. L., Wade, W. H. and Hackerman, N. (1965). *J. phys. Chem.,* **69,** 317.
79 Holmes, H. F., Fuller, E. L. and Secoy, C. H. (1966). *J. phys Chem.,* **70,** 436.
80. Lamond, T. G. and Price, C. R. (1970). *Rubber J.,* **152,** 49.
81. Barry, T. I., Bayliss, R. K. and Lay, L. A. (1968). *J. Mater. Sci.,* **3,** 229.
82. Matijevic, E., Espenscheid, W. F. and Kerker, M. (1963). *J. Colloid Sci.,* **18,** 91.
83. Anderson, P. J. and Livey, D. T. (1961). *Powder Metall.* (7), 189.
84. Johari, O. and Bhattacharyya, S. (1969). *Powder Technology,* **2,** 335.
85. Lippens, B. C., Linsen, B. G. and de Boer, J. H. (1964). *J. Catalysis,* **3,** 32.
86. Lippens, B. C. and de Boer, J. H. (1965). *J. Catalysis,* **4,** 319.
87. Nicolaon G. A. (1969). *J. Chim. phys.,* **66,** 1783.
88. Lippens, B. C. and de Boer, J. H. (1964). *J. Catalysis,* **3,** 44.
89. Marsh, H. and Rand, B. (1970). *J. Colloid Interf. Sci.,* **33,** 478.

90. de Boer, J. H. (1958). The shape of capillaries. *Structure and properties of porous materials,* edited by D. H. Everett and F. S. Stone, Butterworths, London, 68.
91. Wade, W. H. (1964). *J. phys. Chem.,* **68,** 1029.
92. Aristov, B. G., Karnaukhov, A. P. and Kiselev, A. V. (1962). *Russ. J. phys. Chem.,* **36,** 1159, 1348.
Aristov, B. G., Davydov, V. Ya., Karnaukhov, A. P. and Kiselev, A. V., *Ibid.,* 1497.
93. Robinson, E. and Ross, R. A. (1970). *Canadian J. Chem.,* **48,** 2210.
94. Cameron, A. and Stacey, W. O. (1960). *Chem. and Ind.,* 222.
95. Orr, C. (1970). *Powder Technology,* **3,** 117.

CHAPTER 4

SURFACE PROPERTIES

INTRODUCTION

In considering the characterisation of fine powders attention was concentrated on the properties of the primary particles—size, shape, etc. In the next few chapters consideration is given to those uses and associated properties of fine powders where the useful property results from treating the powder—perhaps by dispersing it in a matrix, or sintering. This selection is perhaps a trifle arbitrary, but it enables the important uses of fine powders to be treated separately from what may be termed the more 'chemical engineering' properties.

Any chemical or physical reaction involving a solid occurs at an interface. The mechanism, kinetics, etc. of such reactions can be studied on massive specimens or thin films using suitably sensitive techniques such as field ion microscopy, low energy electron diffraction, etc. The available surface area is small and the concentration of sorbed species or surface reaction product low. By using a fine powder with a large surface:volume ratio these concentrations may be increased, thus facilitating study.

Increased surface area affects the rate of a reaction occurring at the surface. For many reactions such as dissolutions, the rate is approximately proportional to the available surface area. The surface free energy of a fine powder may be appreciable. For iron powder particles a few tens of nm in diameter the surface free energy will be a few hundreds of calories/mol. This represents the free energy associated with defects, departures from stoichiometry, etc., inseparable from the surface, and is usually reduced by adsorption occurring at the surface.

So important is the surface in determining the properties of a fine powder that, at times, it appears that an interface has properties of its own! Thus, in the reinforcement of elastomers, particle size

65

appears more important than the nature of the particle, whether it is carbon, silica, calcium carbonate, etc. Other properties (*e.g.* viscosity control by a suspension of fine silica) are much more closely dependent on the nature of the particle and its surface. In the long run, all properties must be affected to some extent by the nature of the particle and its surface. Such a surface is always covered; the type of bonding between the surface and its covering is dependent on the energetics of the surface which is a reflection of the nature of the particle. This is so whether the particle is exposed to dry inert gas or situated in a metallic or elastomer matrix or in a liquid. When properties appear to be independent of the nature of the particle this is likely to be due to insufficient study or the use of methods which are not sufficiently sensitive.

CHEMICAL NATURE OF THE SURFACE

The overall surface free energy of a solid is reduced by adsorption onto the surface, whether by liquids or gases. To some extent this can be pictured as the utilisation of 'broken bonds' which can be considered to exist because of the discontintuity at the surface of a solid. The strength of the bonding between adsorbate and adsorbent varies continuously from weak physical adsorption of an inert gas, occurring only at low temperature, to chemisorption which commonly has a heat of more than 40 kcal/mol and where bonding is comparable in strength with normal chemical bonding. Desorption of chemisorbed material is often accompanied by chemical changes, so chemisorption shades continuously into chemical reaction. When the bonding is weak, the adsorbate may be replaced by another which bonds more strongly and further reduces the surface free energy. Thus, oxide surfaces, after exposure to air, are covered with layers of water rather than nitrogen. Adsorbent surfaces are not energetically homogeneous because of surface topography and the position of atoms in, and near, the surface. Adsorption sites of differing strengths and reactivities may thus exist so that, on occasions, adsorbate molecules are bonded to specific surface atoms and, on other occasions, the adsorbate may show a certain degree of mobility on the surface.

Most oxides are semi-conducting, due to small departures from stoichiometry. However, in general, it is not necessary to use semi-conductor parameters to describe what is happening on an oxide surface. For example, the effect of the chemisorption of oxygen, hydrogen, etc., on the stoichiometry of zinc oxide is reflected in changes in the conductivity of the powder, changes which can be measured on a fine powder.[1,2]

Since the bonding between surface and adsorbate is a reflection of the nature of the surface itself, all techniques which have been used in such surface studies may be considered to yield information about the surface. One of the most successful of these techniques has been infrared spectroscopy. Others used to advantage include electron spin resonance, nuclear magnetic resonance, Mossbauer spectroscopy, low energy electron diffraction, Auger electron spectroscopy, ESCA and infrared internal reflection spectroscopy.

In the present context an arbitrary limitation has to be imposed on the consideration of adsorption and chemisorption on the surfaces of fine powders. This has been done in such a way as to emphasise those surface modifications which are relevant to the uses of the powders. The vast field of catalysis has been very largely excluded, and only the more important fine powder surfaces—those of silica, alumina, titania, magnesia and carbon—are considered.

Carbon surfaces

The nature of carbon black and the surface groups usually found on it has recently been reviewed,[3] while a somewhat earlier review[4] also covered graphite and diamond and concentrated rather more on methods used for the identification of surface groups. Carbon black is a major industrial product and is available in a variety of particle sizes produced principally either by partial combustion or pyrolysis of highly aromatic refinery by-product oils or by partial combustion of natural gas. In either case, the product contains, in addition to carbon, up to about 4 per cent of oxygen and quantities of both hydrogen and sulphur of less than 1 per cent. Of these impurities, most of the oxygen and about half the sulphur is situated on the surface of the non-porous particles while hydrogen is fairly uniformly distributed over the microcrystallite bundles and dis-organised carbon which form the outer and inner regions, respectively, of the particle.[3] Different blacks contain different amounts of oxygen-containing functional groups on the surface, but phenolic hydroxyl, carboxyl and quinone groups are usually found while, on occasion, there is ample evidence for the presence of lactones and free radicals. These main oxygen-containing groups have been identified by some of the classical reactions of organic functional groups or newly-devised methods.[5]

Sulphur, which is strongly adsorbed on the carbon surface, is removed by heating in vacuum at about 1 000° or in hydrogen at 700°C, partly in the elemental form, and partly as carbon disulphide. It is believed that polysulphide linkages are formed on the carbon surface.[3] The role of these surface groupings in the reinforcement of elastomers containing carbon black is summarised later.

It is interesting to note that infrared spectroscopy, which has been one of the most successful techniques for identifying surface groupings on oxides, is not readily applicable to carbon black because of the very low transmission. Infrared internal reflection spectroscopy is likely to be most useful when further developed.[6]

Other forms of carbon have been less studied, but both graphite and diamond form oxygen-containing groups on the surface. With diamond there is some evidence, from low energy electron diffraction, that distortion of the surface leads to neighbouring atoms approaching each other more closely than usual with some mutual satisfaction of 'broken' bonds. In general, foreign atoms are readily chemisorbed on all forms of carbon, and are then held strongly by covalent-type bonds. Removal of chemisorbed atoms without taking carbon as well is almost impossible and even exchange is difficult.

Silica surfaces

After fairly extensive study, the only group which has been positively identified on the surface of silica is silanol formed by hydroxylation of the surface; the presence of the siloxane group is deduced rather than proved.[4] After the removal of physically adsorbed water at room temperature the fully hydroxylated surface of silica carries about 8 OH/nm^2 (ref. 7).

Commercial silicas vary in their actual degree of hydroxylation, a value of 5–6 groups/nm^2 being quite usual[8] and of these 70 per cent might be adjacent hydroxyls while the remainder are isolated.[9] In the range 170–400°C reversible dehydroxylation occurs and monomeric and dimeric surface silanols are formed. At about 400° less than half of the total hydroxyl has been eliminated. The remaining adjacent hydroxyls form sites for the preferential physical adsorption of water which gives silica its useful dehydrating properties. Above 400°C water is eliminated irreversibly[10] between adjacent hydroxyls, forming siloxane groups so that at 800°C single hydroxyls only remain and the surface is essentially hydrophobic. Some of the siloxane bridges are, in certain circumstances, more reactive towards what are normally regarded as hydrogen sequestring agents, such as aluminium trimethyl, than are single surface hydroxyls.[11] While most of these observations are derived from infrared spectroscopy and thermogravimetric analysis, heat of adsorption measurements lead to similar conclusions and provide some information on the energetics of the hydration processes.[12,13] As the temperature is varied between 400° and 800°C, the concentration of remaining hydroxyls can be controlled. This is important since it gives control over properties, such as thixotropy, which involve the hydroxyl groups. Since silanol groups can undergo various reactions such as chlorination,[14] ammonation[15] and esterification,[16] it is possible to control the surface nature of silica. In this respect esterification or treatment with an alkyl chlorosilane[17] is important since it leads to the formation of a hydrophobic silica. Such a surface leads to improved flow properties since water no longer bonds to the particle surfaces. Improved dispersion in organic media and bonding to organic matrices is to be expected, while thixotropy is reduced. Coatings of this type are thermally stable below about 250°C and the decomposition goes to completion in six hours between 550–750°C. A silanol group reacts as a weak acid, being somewhat more acidic than the carbinol grouping. This concept is useful in comparing the surfaces of silica with other oxides.[4]

Alumina surfaces

Study of the surface of alumina is greatly complicated by the number of crystal modifications which occur and the important part which hydroxyl ions play in the existence of several of these.[20] Under normal conditions, hydroxyl groups cover the surface of alumina and on the γ-form at least three, and possibly five, different types of hydroxyl have been indentified by infrared spectroscopy coupled with dehydration studies.[19] Reactions which occur on the surface of alumina are more diverse than those on silica and show the presence of an electron-abstracting site (thought to result from the formation of an incomplete co-ordination shell round a portion of the aluminium atoms during dehydration[4]), an oxidative

TABLE 14

REACTIONS OF ACTIVE SITES ON ALUMINA

Site	Reaction
Electron abstracting (Lewis acid)	Adsorption of hexachloro acetone to give $CCl_3 \begin{smallmatrix} \\ \end{smallmatrix} C \begin{smallmatrix} =O \\ \end{smallmatrix} CCl_3$, with $O \rightarrow -Al=0$ (22)
Oxidative site	Methyl alcohol oxidation $CH_3OH \rightarrow \cdots \rightarrow Al-O-Al \rightarrow CO$, aluminium formate (23)
Acid-base or ion pair site	Adsorption of carbon dioxide[15,24] Adsorption of 1-butene[15] This site is considered to be a strained oxide link $Al^+O=Al^+$

site and an acid-base site[7] although these sites are not necessarily different. Typical reactions of each type of site are shown in Table 14.

Whilst such reactions as chlorination[15] and ammonation[21] of the hydroxyl groups occur on alumina as on silica, surface compounds on alumina are, in general, more sensitive to water than the corresponding compounds on silica. This is shown by esterification which cannot be used to form a hydrophobic alumina owing to the water-sensitive nature of the esters.[25] However, such a hydrophobic surface can be produced on γ- or α-alumina by reaction with trimethylchlorosilane.[25] From the surface so produced the methyl groups can be burnt off to produce alumina coated with a monolayer of silica.[25]

Titania surfaces

As with other oxide surfaces, the predominant surface grouping is the hydroxyl. The Ti–O bond is more ionic than that between Si–O and this is reflected in the surface reactions of titania. These reactions are not greatly influenced by the crystal form of the oxide, but are more dependent on whether the surface is wet or dry.[26] The dehydration and rehydration reactions of rutile have been studied.[27] Removal of molecular water leads to the production of weak Lewis acid sites while more strongly acidic sites are formed by the removal of isolated hydroxyl groups.[28] These sites probably arise from the existence of titanium ions in the surface in different environments.[29] Chemical evidence also points to the existence of two types of Lewis acid site,[30] both of which can give rise to a series of replacement or adsorption reactions.[31,32] The esterification and conversion to a hydrophobic condition with trimethylsilylchloride occur as with silica and alumina.[25] By reaction of the surface hydroxyls with trialkylaluminium, followed by removal of the alkyl groups by hydrolysis, titania particles may be coated with alumina.[33] Owing to the fairly ready reduction of Ti^{4+} to Ti^{3+}, surface reactions on titania can be quite complicated. As an example of this type of effect, the presence of hydroxyl groups is apparent on the infrared spectrum only after the sample has been exposed to oxygen.[34]

Magnesia surfaces

These have been very little studied, although the presence of hydroxyl on the surface is well known.[35] These hydroxyl groups are quite basic, as might be expected, and form formates with formic acid.[36]

PHYSICAL PROPERTIES OF THE SURFACE

Thixotropy and viscosity control
Suspensions of many approximately equidimensional particles such as grain, coal, ores, cement, etc. in liquids exhibit the so-called Bingham plastic behaviour. This type of behaviour can be interpreted as being due to the formation in the suspension of an interlocking structure able to resist forces less than the yield strength.[37] Such materials may be used to increase viscosity by slurrying, but the relative increase in viscosity is not large. Thus, expressions of the type:

$$\mu_r = \frac{\mu_S}{\mu_L} = 1 + 2\cdot 5V_p + 7\cdot 17V_p{}^2 + 16\cdot 2V_p{}^3$$

(where μ_r is relative viscosity and μ_S and μ_L respectively the viscosity of slurry and liquid and V_p the volume of particulate matter per unit volume of liquid) have been used up to quite high concentrations of solids.[38] Such relationships do not have a particle size term.

Thixotropy is a decrease in viscosity with continuing shear, and is often exhibited by suspensions of anisometric particles—discs, plates, etc. Thixotropy has been considered as the result of particles tending to aggregate or form definite structural linkages. One possible mechanism for this is hydrogen bonding.

The predominant surface grouping on silica, alumina or titania which has been exposed to the air is hydroxyl. Hydroxyl groups tend to form hydrogen bonds to other hydroxyl groups and it may be surmised that the existence of such bonds is, in part, responsible for the high degree of agglomeration of fine powders. Thus, when a fine powder of this hydroxylated type is immersed in a liquid its behaviour will depend on the polarity of the liquid. In a polar liquid which is itself capable of hydrogen bonding, the powder particles will be surrounded by, and will bond to, the liquid. In a non-polar liquid after dispersion the particles will tend to bond to each other again.

Particle–particle bonding causes a much greater increase of viscosity than does bonding of particle to the liquid medium. Both types of bonding are easily broken by shear and reform on standing.

To compare these different types of hydrogen bond, Aerosil forms a gel in *n*-thiobutanol at 4·6 g/100 ml, while, in the polar *n*-butanol, 11·6 g/100 ml is needed to gel.[39] A silica which has a lower concentration of surface hydroxyl groups (because of particle hydrophobisation by esterification, etc.) shows a markedly smaller viscosity increase when compared with a fully hydroxylated silica in a nonpolar solvent.[40]

This increase of viscosity, which can occur when the correct amount of a finely-divided hydroxylated or ammonia-covered powder is added to either a polar or a non-polar liquid, is thus a function of the degree of hydroxylation of the surface, the fineness of the powder and the water (or ammonia) content of the powder prior to addition. Further, in principle, higher viscosity and greater thixotropy can be obtained in both polar and non-polar liquids by the use of suitable additives. In non-polar or low polarity liquids the addition of 10–20 per cent by weight (based on silica) of water, ethylene glycol or other substance capable of multiple hydrogen bonding forms a tightly knit silica chain structure. In polar systems a similar addition, but of a long chain polar substance such as *n*-octylamine, acts similarly.[41,42] Figure 19 shows how the interparticle bonding is formed.

At the same time as the viscosity is increased by hydrogen bonding through the liquid, it is also stabilised against temperature change. Although fine silica is usually used for the control of viscosity, fine alumina may also be used and may be preferable in aqueous systems at higher pH. As well as the more straightforward thickening and viscosity control, the addition of these fine powders leads to reduced settling of suspended matter, whether solid or liquid. The following are examples of types of product where these properties are used. In adhesives and sealants 'sag' is decreased and better adhesion obtained.[43] In surface coatings such as inks, varnishes and paints, settling of pigments is reduced and increased viscosity and thixotropy may make for easier application.[41,43−46] In lubricants the stability of viscosity with temperature is of value; in pharmaceuticals viscosity increase enables greases and ointments to be made and reduces settling and the need to 'shake the bottle'.[47] Fine powders increase the settling time for aerosols and can lead to improved efficiency and uniformity of spread.

For all applications involving surface bondings, non-porous powders are likely to be more valuable than those with pores;

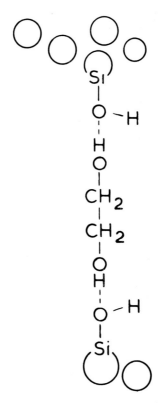

Fig. 19.   *Hydrogen bonding between silica particles via ethylene glycol*
additive.

hydroxyl groups situated in pores may not be available for hydrogen
bonding. While there does not appear to be any quantitative data,
if the proposed mechanism is correct there should be a correlation
between the concentration of hydroxyl groups on the surface of the
powder and its surface area on the one hand and the increase of
viscosity in a given system on the other.

Somewhat related to the usages mentioned above are those in
which a liquid is completely adsorbed on a fine powder (*see* page 138).

Zeta potentials

Colloidal particles acquire a charge when placed in an electrolyte
solution or polar medium. This is usually considered to be the result
of selective adsorption of one type of ion on the surface of the

colloidal particle or selective dissolution of one type of ion from the colloidal particle and can be seen in the sign and magnitude of the zeta potential. This zeta potential is the potential due to the concentration of counter ions in the solution around the colloidal particle in order to achieve overall neutrality of the solution[48] (*see* Fig. 20).

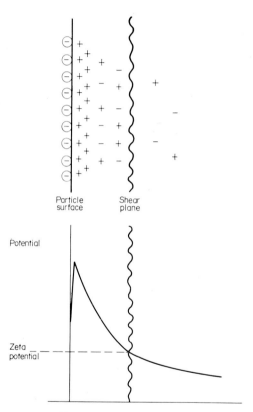

Fig. 20.   *Origin of the zeta potential.*

The magnitude and sign of the zeta potential is therefore a reflection of the nature and extent of surface adsorption onto the particle. Thus, because of the acidic nature of the hydroxyl groups on alumina[4] a suspension in an aqueous electrolyte solution is positively charged at pH 9.[49] Silica, on the other hand, shows a negative charge at pH as low as 4·5[47] while the isoelectric point for titania is at pH 6·6.[50] Use is made of the charge on alumina to reduce mutual repulsion between negatively charged paper fibres and to assist in anchoring fillers.[51] It has been suggested that similar use can be made of alumina with wool, cotton and other negatively charged fibrous materials.[49]

The positive charge on alumina particles in a polar medium may also be useful on occasion as a flocculating agent for negatively charged dispersions.

Electroviscous effects

This term is applied to the changes in the viscosity of suspensions owing to the presence of electric charges, either spontaneous or imposed by a high potential. In order that the effects may be appreciable, the particles must be small.

In electrolyte solutions electrical double layers are formed round the charged particles. When the particles are, on the average, well separated, the only effect of the imposition of shear will be some distortion of the double layer. This distortion produces an electrostatic contribution to the viscosity and this very small effect is known as the first electroviscous effect. At higher concentrations of suspension there will be some interaction of double layers[52] during shear. The magnitude of this second electroviscous effect increases as the square of the particle concentration and also increases as the ionic strength of the solution is reduced since the double layer is then thicker.[53,54] The effect is greater when the thickness of the double layer is greater than the particle radius[55] but even then may only result in something like a doubling of the viscosity[53] when a few per cent by volume of suspended matter is present. Under conditions of shear such a suspension will exhibit non-Newtonian behaviour with a certain amount of thixotropy.[53] The thixotropic behaviour, described earlier and ascribed to hydrogen bonding, is much greater than that which might result from these electroviscous effects.

In still more concentrated suspensions (40–50 volume per cent) something approaching close packing of particles with their double layers should occur[56] at least in the presence of metallic soaps, etc., which might stabilise the suspension. Application of a high electrical potential (~ 30 kV/cm) across a thin layer of such a suspension in a semi-insulating medium produces a much greater increase in viscosity than the increases due to the above-mentioned electroviscous effects. Shear resistance may reach several hundreds of grams/cm^2 across a thin layer of such a suspension[57] and, as distinct from the more usual thixotropic effect, this resistance is maintained when shear actually occurs. The effect has been pictured (Fig. 21) as a polarisation of the double layer round each particle. Overlap of the double layers of adjacent particles then occurs with greatly increased interaction so that additional energy must be dissipated on shear normal to the direction of the applied field.[56,58] To form the suspension a wide range of materials has proved suitable —alumina,[59] boron,[60] activated charcoal, clay, mica, mannitol, lead

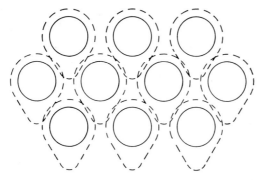

Fig. 21. The effect of an applied electric field causing overlap of the electrical double layers surrounding particles in an electroviscous fluid.[58]

oxide, barium titanate and others,[61] in the size range about 0·05–10 μm but more usually round 1 μm. The most used material is silica, probably because so much more is known about its surface chemistry than that of the other materials. The optimum effect is obtained when the silica is almost fully hydroxylated (8 silanol groups/nm^2) and carries relatively little (1–2 molecules/nm^2) physically adsorbed water.[62,63] The significance of these values in the context of the mechanism of the electroviscous effect is not clear. In general, the suspended solid should have a high surface conductivity with respect to the conductivity of the oil in which the suspension is to be made. It is valuable if the particles possess the ability to adsorb water or alcohols.[61] Each particle can then be pictured as coated with an insulating oil layer beneath which there is a partially conducting surface over which the charge may move. A surface active agent lubricates the boundaries between the particle regions. It is often necessary to add further surface active agent to reduce the viscosity of the suspension, or, if a grease is required, viscosity may be increased by adding an acrylic derivative.[61] Although the applied potential is high, the electrical resistance of the suspension is such that only a few milliamps current flows.[57]

A number of suggestions have been made for the utilisation of this effect, including electrically-activated switches, clutches, valves, brakes and chucks.[64-67]

At present, the bulk of the literature concerning electroviscous fluids exists in patents. The recipes given are quite complicated, but the mechanism by which these fluids undergo such rapid and extreme changes of viscosity is not known. Because of the technological stimulus, it seems probable that the development of useful devices, such as brakes, clutches, switches, etc., using electroviscous fluids, will precede any elucidation of the way in which the fluids function. As a prelude to study of the latter, some *ad hoc* comparisons to show the effect of changes of particle size, shape, bulk chemical composition and surface chemistry of the particulate phase, coupled with changes in the nature of the fluid, surface active additives, etc., may suggest useful lines of attack, as well as providing information for improving the recipes. It is not clear how much can be done to set up a simplified model system for the study of the mechanism. The whole field of electroviscous fluids appears to need complete and careful study and the rewards in terms of useful devices could well be considerable.

Catalysis

No account of the properties and uses of fine powders would be complete without some mention of the field of catalysis. However, there is such variety in the catalysts used and the types of reaction which may be catalysed that it is quite impossible to do more than generalise in the most superficial terms.

Many of the catalysts currently in use are microporous with relatively large internal surface area. In these materials large surface area can be achieved without the handling difficulties which attend the use of very small particles. Catalysts may be of one component, or may consist of a second component deposited on a support. In the latter case there may be a subtle interplay between the two components. Many catalysts function by forming some transitory activated species following adsorption from the vapour phase on some activated sites on the catalyst. The catalyst possesses a large internal surface area, and many of these pores are less than a few tens of nanometres in diameter. Diffusion into and along such pores will be fairly slow so that fast reactions will use only the mouths of the pores while slower reactions may proceed throughout the body of the pores. By relating these factors to gas flow rates, a certain degree of selectivity can be achieved. However, when gas flow rates are high and reaction is rapid, only a small proportion of the total surface area of a porous catalyst may, in fact, be used. Under these conditions the use of a non-porous, high surface area catalyst may be advantageous.

The study of catalysis is essentially a study of the surface chemistry and energetics of the catalyst, its pore structure and size distribution, and its interaction with a gaseous environment. Among techniques which have been used to prepare porous catalysts the most important groups are:

—preparation and drying of a gel;
—leaching of one component of a multicomponent body;
—thermal decomposition.

Before the catalytic behaviour of a given substance can be understood, an extensive study by a wide range of techniques such as infrared, nuclear magnetic resonance and electron spin resonance spectroscopies and others is needed to learn something of the transitory adsorbed species in addition to complete characterisation of the catalyst. An example is found in a review, 'Organic catalysis over crystalline alumino silicates'.[68]

For a discussion of particular aspects of catalysis reference may be made to standard works (for example, reference 69) or to relevant series of reviews.[70,71]

REFERENCES

1. Narayana, D., Subrahma, V. S., Lal, J., Ali, M. M. and Kesavulu, V. (1970). *J. phys. Chem.,* **74,** 779.
2. Guillen, R., Palz, W. and Yvroud, E. (1970). *Compt rendu* (B), **270,** 101.
3. Deviney, M. L. (1969). *Adv. Colloid Interf. Sci.,* **2,** 237.
4. Boehm, H. P. (1966). Chemical identification of surface groups. *Adv. Catalysis,* **16,** edited by D. D. Eley, H. Pines and P. B. Weisz, Academic Press, New York, 179.
5. Given, P. H. and Hill, L. W. (1969). *Carbon,* **7,** 649.
6. Mattson, J. S., Mark, H. B. and Weber, W. J. (1969). *Anal. Chem.,* **41,** 355.

7. Hair, M. L. (1967). *Infrared spectroscopy in surface chemistry*, Arnold, London.
8. Bode, R., Ferch, H. and Fratzscher, H. (1967). *Kaut. Gummi Kunstst*, **20**, 578.
9. Armstead, C. G., Tyler, A. J. Hambleton, F. H., Mitchell, S. A. and Hockey, J. A. (1969). *J. phys. Chem.*, **73**, 3947.
10. Young, G. J. (1958). *J. Colloid Sci.*, **13**, 67.
11. Kunawicz, J., Jones, P. and Hockey, J. A. (1971). *Trans. Farad. Soc.*, **67**, 848.
12. Whalen, J. W. (1961). *Adv. Chem.*, **33**, 281.
13. Young, G. J. and Bursh, T. P. (1960). *J. Colloid Interf. Chem.*, **15**, 361.
14. Folman, M. (1961). *Trans. Faraday Soc.*, **57**, 2000.
15. Peri, J. B. (1966). *J. phys. Chem.*, **70**, 2937, 3168.
16. Folman, M. and Yates, D. J. C. (1958) *Proc. roy. Soc. A*, **246**, 32.
17. Eakins, W. J. (1968). *Indus. Engg Chem: Product Res. Dev.*, **7**, 39.
18. Zhuravlev, L. T., Kiselev, A. V. and Naidina, V. P. (1968). *Russ. J. phys. Chem.*, **42**, 1200.
19. Peri, J. B. and Hannan, R. B. (1960). *J. phys. Chem.*, **64**, 1526.
20. Glemser, O. and Rieck, G. (1956). *Angew. Chem.*, **68**, 182; (1958). *Zeit. anorg. Chem.*, **297**, 175.
21. Peri, J. B. (1965). *J. phys. Chem.*, **69**, 220, 231.
22. Hair, M. L. and Chapman, I. D. *Ibid.*, 3949.
23. Greenler, R. G. (1962). *J. chem. Phys.*, **37**, 2094.
24. Little, L. H. and Amberg, C. H. (1962). *Canad. J. Chem.*, **40**, 1997.
25. Stober, W., Lieflander, M. and Bohn, E. (1960). *Beitrage zur Silikose-Forschung Special Vol.* 4, Bergbau-Berufsgenossenschaft, Bochum, 111.
26. Primet, M., Basset, J., Mathieu, M-V. and Prettre, M. (1970). *J. phys. Chem.*, **74**, 2868.
27. Jackson, P. and Parfitt, G. D. (1971). *Trans. Farad. Soc.*, **67**, 2469.
28. Primet, M., Pichat, P. and Mathieu, M-V. (1971). *J. phys. Chem.*, **75**, 1221.
29. Parfitt, G. D., Ramsbotham, J. and Rochester, C. H. (1971). *Trans. Farad. Soc.*, **67**, 841.
30. Herrmann, M. and Boehm, H. P. (1969). *Z. anorg. Chem.*, **368**, 73.
31. Flaig-Baumann, R., Herrmann, M. and Boehm, H. P. (1970). *Z. anorg. Chem.*, **372**, 296.
32. Herrmann, M., Kaluza, U. and Boehm, H. P. (1970). *Z. anorg. Chem.*, **372**, 308.
33. Lieflander, M. and Stober, W. (1960). *Zeit. Naturf.*, **15B**, 411.
34. Lewis, K. E. and Parfitt, G. D. (1966). *Trans. Farad. Soc.*, **62**, 204.
35. Anderson, P. J., Horlock, R. F. and Oliver, J. F. (1965). *Trans. Faraday Soc.*, **61**, 2754.
36. Scholten, J. J. F., Mars, P., Menon, P. G. and Van Hardeveld, R. (1965). *Proc. 3rd Int. Congr. catalysis,* North Holland Publishing Co., Amsterdam, 881.
37. Orr, C. (1966). *Particulate technology*, Macmillan, New York, 127.
38. Vand, V. (1948). *J. phys. Chem.*, **52**, 300.
39. Brunner, H. (1960). *Informationsdienst of Study Group for Pharmaceutical Technology*, Mainz, (3), 42.
40. Bode, R., Ferch, H. and Fratzscher, H. (1968). *Paint Oil Colour J.*, **154**, 415.
41. Schue, G. K. (1968). *Amer. Paint J.*, **52**, (55), 16.
42. Monsanto Ltd. British Patent, 1,200,745 (29 July, 1970).
43. Sweeney, T. R. (1967). *Adhes. Age*, **10**, 32.
44. Elbrachter, A. (1968). *Paint Varn. Prod.*, **58**, (6), 63.
45. Fratzscher, H. (1968). *Double Liaison* (152), 469.

46. Schue, G. K. (1969). *S.P.E. Jl.*, **25**, (7), 40.
47. *Cabosil, how to use it, where to use it,* Cabot Corporation, Boston, Mass., 1968.
48. See for example Adamson, A. W. (1967). *Physical chemistry of surfaces,* 2nd edition, Interscience, New York, 209.
49. *Alon fumed alumina.* (1968). Cabot Corporation, Boston, Massachusetts.
50. Boehm, H. P. (1966). *Angew. Chem. Int. Ed.,* **5**, 533.
51. Cameron, S. S. (1961). Powders in the paper industry. *Powders in industry,* Society for Chemical Industry, London, 323.
52. Conway, B. E. and Dobry-Duclaux, A. (1960). Viscosity of suspensions of electrically charged particles and solutions of polymeric electrolytes. *Rheology,* **3**, edited by F. R. Eirich, Academic Press, New York, 83.
53. Harmsen, G. J., Schooten, J. V. and Overbeek, J. Th. G. (1953). *J. Colloid Sci.,* **8**, 64, 72.
54. Dobry, A. (1955). *J. Chim. physique,* **52**, 809.
55. Stone-Masui, J. and Watillon, A. (1968). *J. Colloid Interf. Sci.,* **28**, 187.
56. Martinek, T. W. and Klass, D. L. (1965). *Nat. Lubricating Grease Inst. Spokesman,* **29**, 219; Martinek, T. W. Haines, R. M., and Klass, D. L. (1966). *Ibid.,* **30**, 286.
57. Winslow, W. M. (1949). *J. appl. Phys.,* **20**, 1137.
58. Klass, D. L. and Martinek, T. W. (1967). *J. appl. Phys.,* **38**, 67, 75.
59. Martinek, T. W. and Klass, D. L. U.S. Pat. 3,367,872 (6 February, 1968).
60. Martinek, T. W. and Klass, D. L. U.S. Pat. 3,399,145 (27 August, 1968).
61. Winslow, W. M. U.S. Pat. 3,047,057 (31 July, 1962).
62. Martinek, T. W. and Klass, D. L. U.S. Pat. 3,250,726 (10 May, 1966).
63. Martinek, T. W. U.S. Pat. 3,397,147 (13 August, 1968).
64. Klass, D. L. and Brozowski, V. U.S. Pat. 3,385,793 (28 May, 1968).
65. Winslow, W. M. U.S. Pat. 2,661,825 (8 December, 1953).
66. Winslow, W. M. U.S. Pat. 2,663,809 (22 December, 1953).
67. Winslow, W. M. U.S. Pat. 2,886,151 (12 May, 1959).
68. Venuto, P. B. and Landis, P. S. (1968). Organic catalysis over crystalline aluminosilicates. *Adv. Catalysis,* **18**, edited by D. D. Eley, H. Pines and P. B. Weisz, Academic Press, New York, 259.
69. *Catalysis* in seven volumes edited by P. H. Emmett, Reinhold, New York, 1954–60.
70. Annual *Advances in catalysis,* Academic Press, New York, Now reached **21** (1971).
71. A new series *Catalysis reviews,* Dekker, London, of which **4** appeared in 1971.

CHAPTER 5

FINE POWDERS IN REINFORCEMENT

INTRODUCTION

Dispersions of fine particles of a second phase in a matrix of a first phase are quite widely used to strengthen the first phase. The nature of the dispersed phase varies with that of the matrix and combinations which may increase strength are shown in Table 15. Since different types of matrices fail mechanically in different ways, it is to be expected that the mechanisms of strengthening will vary, and this is certainly true. Many details of the proposed mechanisms are, at the moment, subjects of considerable discussion.

Of the types of strengthening listed in Table 15, precipitation strengthening is unique in that, although the strengthening precipitate consists of fine particles, these are generated *in situ* and do not have an existence separate from the bulk metallic matrix. The pertinence of mentioning such an application in a chapter covering the usages of fine powders is thus somewhat doubtful and it is included only for the sake of completeness.

TABLE 15

MATRIX–DISPERSED PHASE COMBINATIONS LEADING TO
STRENGTHENING

Nature of matrix phase	*Nature of dispersed reinforcing phase*	*Type of strengthening*
Metallic	Metallic	Precipitation strengthening
Metallic	Hard ceramic	Dispersion strengthening
Ceramic	Metallic	Dispersion strengthening
Elastomer	Inorganic or carbon	Reinforcement
Metallic	Hard ceramic	Cermet or cemented ceramic (Higher proportion of refractory phase than in dispersion hardening)

PRECIPITATION STRENGTHENING

In precipitation strengthening the second phase is generated within the grain of the metallic matrix. A supersaturated solid solution is produced by quenching from above the homogenisation temperature; annealing then forms precipitates containing 100–1 000 atoms. Conditions for the formation of the precipitate vary with the system, being governed by the relative solubilities of the components and the ability of the precipitate to nucleate. The finely dispersed phase will redissolve at high enough temperatures, so precipitation-strengthened metals can be used only at low temperatures.

Precipitation strengthening involves differences between precipitate and matrix (*e.g.* in atomic volume if the interfaces are coherent and elastic strain fields are appreciable, or in stacking fault energy) as well as the increased stress required to force the dislocation between the obstacles which are precipitate particles.[1–3] These are factors having some parallels in dispersion strengthening.

DISPERSION-STRENGTHENED METALS

In dispersion-strengthened metals the dispersed phase is a hard ceramic. As a result, methods of preparation and fabrication are quite different from those for precipitation strengthening. The metallic matrix cannot now be fused without the risk of destroying the dispersion (*see* below) and fabrication, joining, etc., become more difficult. An overall impression of the whole field of dispersion-strengthened metals may be obtained from several reviews.[4–6]

The prototype of modern dispersion-strengthened metals (sintered aluminium powder with oxide dispersion) was announced in 1950. Since that time, a very wide range of metallic matrices and of finely-divided hard particles—usually oxides—has been studied. In spite of this wealth of data, there is no completely accepted view of the detailed mechanism of strengthening by hard dispersions. A critical examination of theories for the yield strength, work hardening and creep has shown that, whilst a number of the theories are of value, it is probable that no one theory will successfully describe the behaviour of all dispersion-strengthened alloys.[7,8]

In general, dispersion-strengthened metals are weaker than conventional alloys at lower temperatures, but more stable and stronger at higher temperatures. A comparison of rupture strengths is shown in Fig. 22.[9]

Stored energy resulting from cold working during fabrication appears to be responsible for much of the increase in, for example, rupture strength.[4] The dispersed phase hinders dislocation movement so that recrystallisation is greatly delayed (often to temperatures not far from the melting point) and this stored energy is retained.

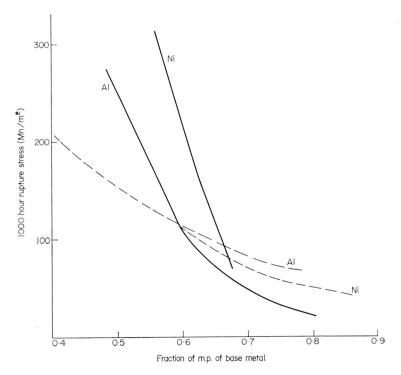

Fig. 22.   Rupture strength of typical dispersion-strengthened (dotted line) and wrought alloys (full line).[9]

Properties of the dispersion

(a) *Particle size and shape:* Experiments suggest that, within the size range studied (*i.e.* above about 5 nm) the finer the dispersion, the greater the strength. Very little is known about the optimum shape of dispersed particles[4] or optimum inherent strength.[10]

(*b*) *Particle spacing:* Several studies have shown the importance of particle spacing (usually measured by the volume fraction of dispersed phase), but no optimum value has been suggested.[11,12] This is probably not surprising because of the widely different conditions at the interface between a variety of dispersed phases set in a variety of matrices. Although most of the theories suggest that a uniform dispersion of hard particles in the matrix is required, and the achievement of such a dispersion has been the aim of most studies, it has recently been shown that a network of alumina particles distributed along cell walls in aluminium, *i.e.* as a network, shows values of tensile strength, both at room temperature and 400°C, similar to those for a uniform dispersion.[13]

(*c*) *Particle–matrix interface:* The interface is important in several respects. Its degree of coherence can affect the mechanical properties; its thermal stability can affect the temperature at which the composite is usable. Very small particles, especially those generated by internal oxidation or nitridation (*see* page 87) may be coherent, while larger particles in the same combination become only partly coherent and strain fields introduced into the matrix are reduced and strength may fall.[14,15] For example, in internally oxidised copper alloys voids form in fracture first on larger particles and size dependence is shifted to smaller size with decreasing temperature.[16] To achieve strengthening, then, the matrix should wet the dispersed particles[17] but there must be no chemical reaction, nor, at the temperature of use, must coarsening be appreciable. Such coarsening occurs by diffusion through the matrix not only of the dispersed phase itself but also of the dissociation products which exist in equilibrium with the dispersed phase.[18] The driving force for coagulation is the minimisation of the surface free energy, and since, in general, the oxide–metal interface has a lower free energy than metal–metal combinations, oxides coarsen less than most strengthening metallic precipitates.[15] Impurities which segregate to the interface can exert an important effect on coarsening. For example, iron or sulphur increase the rate of coagulation of thoria dispersed in nickel.[19] Phase changes such as $\gamma \rightarrow \alpha$ alumina also accelerate coagulation[20] and thus reduce the impedence to dislocation motion and so to recrystallisation. For reasons of stability and chemical inertia, oxides have been favoured as the dispersed phase, especially those with a high heat of formation and without phase transitions or sub-oxides. However, even when all these criteria are satisfied there may be some coarsening. For example, careful microscopic examination shows that the corners of thoria particles dispersed in a matrix of tungsten are gradually rounded.[21] The recent development of a method for making direct measurement of the interface strength between dispersed phase and matrix by rapid temperature cycling and determination of decohesion[22] can be expected to provide useful data for the interpretation of the mechanical properties of many dispersion-strengthened alloys.

Production of suitable dispersions in metallic matrices

Methods for the production of oxide dispersions have been reviewed:[23,24]

(a) *Mechanical blending:* This is basically a straightforward method which has been used with many different components.[25,26] There are, however, drawbacks. Blending may lead to agglomeration which remains undetected; density differences between the phases may lead to incomplete blending.[27] To prevent reagglomeration, grinding aids (oleic acid in heptane) may be added and the fine metal cleaned in hydrogen after loose compaction.[28] Very recently, mechanical blending using a high energy driven ball mill has been introduced. The balls and charge are agitated by impellers.[29] Grinding and blending occur simultaneously.

(b) *Preferential reduction of the more noble component in a mixture of fine oxides:* Thoria and alumina are typical non-reducible hard oxides and nickel, iron, chromium and molybdenum or combinations of these are typical matrix materials.[19,30,31] The necessary uniform intimate mixture of fine oxides can be prepared by coprecipitation of hydrated oxides, oxalates, etc. followed by suitable heat treatment. The reduction equilibrium must be such that the reaction goes strictly to completion.

(c) *Salt decomposition onto metal surfaces:* Coating of a metal powder with a suitable solution, followed by thermal decomposition of the salt deposited from solution, can give a uniform coating, especially of oxides.[32,33]

(d) *Vapour deposition:* This technique, which has not been used as much as might have been expected, would appear to be very flexible, so that coatings of a metallic matrix may be deposited on the hard dispersed phase[25,34] or a coating of dispersed phase on the powdered matrix.

(e) *Electrodeposition:* Submicrometre sized particles of alumina and of titania suspended in a nickel plating bath have been codeposited with the metal. Stirring and ultrasonic vibration were used to prevent settling of the oxide suspension and care was needed to avoid the presence of quantities of electrolyte sufficient to cause flocculation. The electrodeposit contained the oxide phase at the spacings and loadings usually used for dispersion strengthening, but such high temperature strengthening was not apparent until suitable thermomechanical treatment had been given.[35]

(f) *Surface oxidation (nitridation, etc.):* By heating a flake or powdered metal matrix under suitable conditions, a thin protective film of oxide (nitride, etc.) is formed. During the compaction and extrusion needed to convert any powdered material into the massive form, the coating is broken and a fine dispersion obtained. Two commercial alloys, dispersion-strengthened aluminium[36] and dispersion-strengthened lead,[37] both containing dispersed oxides, are made in this way.

(g) *Internal oxidation (nitridation, etc.):* This process is applicable to a dilute alloy in the form of sheet, wire or powder when the solute can be oxidised (nitrided, etc.) preferentially. Diffusion rates must be high and that of the gas phase (oxygen, nitrogen) into the specimen greater than that of the solute to the surface. Magnesium and aluminium oxides dispersed in silver,[38] titanium carbide in nickel[39] several nitrides in a molybdenum matrix[40] and nitrides in a matrix of austenitic steel[41] have been prepared in this way.

(h) *Comparison of preparative methods:* Because of the many variables involved a quantitative comparison of the strengths of materials made by these processes while possible is not easy. A comparison of direct mixing, co-precipitation of hydroxides followed by ignition to oxides, and direct mixing of oxides with oxide then reduced to metal was made[42] after sintering and extruding. Oxide reduction appeared to give the strongest product, but the dispersion was much larger than the particle size of the starting oxide. Another study along similar lines but with alumina in a cobalt-nickel alloy found that selective reduction gave a more uniform dispersion than direct mixing and grain growth was less. The alumina particles in this instance were about 20 nm in diameter.[43] In general mixing behaviour will vary from one composition to another; it is believed that direct blending is sometimes capable of producing material which is as strong as that prepared by more sophisticated methods.

The outlook for dispersion strengthening

If no clear picture of the present status and future outlook for dispersion-strengthened metals emerges from the brief comments above, this is, in part at least, due to confusion existing in the field. More than twenty years have now elapsed since the announcement of dispersion-strengthened aluminium (SAP), yet there are very few available commercial materials, and those are relatively little used. Nickel and nickel-chromium alloy are available with thoria dispersions and lead containing a lead oxide dispersion and thoria in tungsten are also available, but few others. There appear to be at least three factors bearing on the failure of dispersion-strengthened metals to achieve much penetration of the market:

—cost;
—properties of the materials;
—the complex interactions governing selection of a particular alloy system.

Sufficient has been said above concerning the methods available for the production of dispersion-strengthened metals to show that such methods are not simple. Apart from the surface oxidation procedure applicable to SAP, most of the preparative routes are unduly slow (*e.g.* internal oxidation depending on diffusion rates) or complicated (*e.g.* selective reduction). In addition, it is necessary to use a variety of thermomechanical processes to introduce the desired high temperature strength. Materials made by such routes will never be cheap, although present costs will probably be reduced. Ideally, the preparative route should be via the melting pot so that large-scale production can be considered. Efforts have been made to produce alumina-aluminium by such a route,[44] but attempts to disperse a powder in a molten metal by ultrasonic vibration[45] did not produce the required uniform dispersion.

Another result of the instability of the dispersion at high temperatures is that conventional forming and joining techniques cannot be used. Cladding or diffusion bonding seem more appropriate. This is another factor which discourages use.

It is probable that dispersion-strengthened materials will be considered whenever it is required to extend the use of a particular alloy or matrix to higher temperatures than can be achieved with conventional strengthening. In this way, use may still be made, for example, of the corrosion resistance of aluminium or the thermal conductivity of copper.[9] When there is no special reason for continuing to use a particular matrix, high temperature strength may be achieved by changing to a different system progressively along the series:

$$Ni \rightarrow Cr \rightarrow Nb \rightarrow Mo \rightarrow Ta \rightarrow W$$

However, the most refractory metals oxidise very readily and can be used in air only when coated. Other materials which might be used under these conditions, e.g. intermetallics and ceramics, are brittle. There is, therefore, a need for a strong ductile oxidation-resistant material for use in the temperature range around 1 200–1 400°C (at, and just beyond, the limits of current superalloys). Addition of chromium to a nickel-base alloy improves oxidation resistance; addition of a thoria dispersion to a nickel-chrome alloy further improves oxidation resistance.[46] However, there are indications that the addition of chromium leads to reduced stability of the thoria dispersion[19] and also that high temperature strength is not as good as anticipated.[47]

It is now becoming apparent that, in some instances, it is possible to combine dispersion strengthening with precipitation strengthening and, in this way, to retain the best features of both mechanisms.[29] There is, of course, a very wide range of possible disperse phases besides oxides although there are suggestions that the latter are the most stable.[39] Interaction between the dispersion and precipitate or mismatch in the thermomechanical treatments required to achieve high strength due to dispersion and precipitation hardening may prevent general realisation of the optimum strength due to both mechanisms. First reports cover a nickel 20 per cent chromium alloy hardened with aluminium and titanium and with alumina and yttria as dispersed phases. At lower temperatures the rupture strength behaviour is that of ordinary nickel-based age-hardened alloys while, at higher temperature, dispersion hardening takes over.[29]

Many future developments may be expected along these lines, and, if successful, much more use will probably be made of dispersion strengthening as materials are required to withstand more and more extreme conditions, especially as great improvements in high temperature strengths by conventional alloying appear somewhat unlikely.

Ductility introduced by a dispersion

Whilst ductility is normally reduced by introducing a dispersion,[9] under certain conditions it appears that the temperature of the ductile-brittle transition is lowered and cleavage stress is increased (equivalent to increased ductility). This occurs, for example, with a dispersion of thoria in an iron matrix.[48] It is believed that, as in a ductile matrix, the dispersed phase blocks dislocation movement. This limits the pile-up length and thus, in a brittle matrix, reduces the potency of pile-up as a crack-inducing stress concentrator. Alternatively, the particles may act as sources of dislocations and thus promote stress relaxation by slip and make cleavage more difficult. A complex relationship may be expected between particle size and ductility if the first of these mechanisms predominates, but there is no experimental data. At present, a particle size of 25–100 nm is considered desirable.[49] Increased ductility in tungsten[48] and chromium[50] matrices has been reported, and work on other systems is probably in hand.

DISPERSION-STRENGTHENED CERAMICS

Fine molybdenum dispersed in an alumina matrix inhibits grain growth. Increased strength in the ceramic matrix results from the finer grain size on the basis of the Knudsen or other similar relation. Starting with submicrometre molybdenum and alumina, matrix grain size less than 2 μm and 98 per cent relative density can be obtained after consolidation. Fracture energy is about 50 per cent greater than that of alumina of the same grain size.[51] Cutting tools of such strengthened alumina are expected to have lives up to five times as long as normal alumina tips.[52,53] This method of controlling grain growth is similar in action to the use of non-metallic grain growth inhibitors which are quite widely used in sintering ceramics and which appear to operate by deposition in, and pinning of, grain boundaries (*e.g.* reference 54).

A dispersion of fine tungsten in uranium and plutonium carbides controls fission gas swelling. The tungsten was precipitated by annealing at 1 400°C to exsolve it from the carbide into which it had been introduced by zone refining.[55]

Dispersion strengthening would appear to provide a means of retaining a fine grain size in a ceramic without the need for the complexities and restrictions inherent in hot pressing. Whether the presence of the dispersion can be tolerated must depend on the use envisaged for the product.

REINFORCEMENT OF ELASTOMERS

Elastomers are composed of a tangled mass of kinked and intertwined chain-like molecules which, above the glass transition temperature, are in a state of constant thermal motion. Elasticity arises from resistance to forces tending to distort the normal pattern within the solid. The tangled chains are free to slide past each other except where they are tied together by cross links. Such links limit the amount of stretch and increase the elasticity, and are formed by reaction of the chains with certain additives (curing agents or accelerators) mixed into the mass. Heating also increases the extent and rate of cross linking and is termed vulcanisation.[56] Natural and synthetic elastomers vary greatly in their properties but, in general, are not able, as they stand, to meet the more stringent demands placed upon them, and need to be reinforced.

Depending on the function being fulfilled, elastomers deteriorate in two main ways.[57] When used for energy absorption (as in cushioning, silencing, etc.) repeated deformations, small relative to the ultimate breaking strength, lead to a form of fatigue and a deterioration in the hysteresis behaviour which is the main method of dissipating energy. In other types of usage (e.g. road tyres) failure occurs by abrasion in which the surface of the rubber is torn away. There have, therefore, been two approaches to the study of the strength of rubber, one via the hysteresis, the other via abrasion due to cut growth, explained by the concept of tearing energy. Recently, attempts have been made to find common ground in these two approaches.[58] For amorphous rubbers there is a relationship between strain at break and hysteresis at break[59] so tensile strength is often used as a measure of usable strength; like tear failure, tensile failure is initiated by some stress raising discontinuity. Tensile strength is always higher than tear strength as the volumes subjected to stress concentrations which initiate failure are smaller in tensile failure. However, under certain conditions, for instance when the tear initiates at a notch, tear strength and tensile strength may be widely different. To resist abrasion, stiffness and mechanical hysteresis are needed, as well as good tear strength. Stiffness prevents excessive elongation whilst mechanical hysteresis leads to energy loss as heat, and consequent reduced wear.[57]

It is difficult exactly to define reinforcement in terms of a single property of the elastomer system. For example, reinforcement is often considered to be represented by an increase in tensile strength, but, in some natural rubbers which already possess good tensile strength, such increase may be quite marginal. It is probably preferable to consider as reinforcing a filler which causes increased stiffness, elastic modulus and mechanical hysteresis, though usually at the expense of resilience and elasticity.[56]

Within this definition and within reason it is true to say that any hard finely-divided solid of about 20–40 nm particle size will reinforce rubber.[60] A number of recent reviews cover, in much greater depth than is justified here, many aspects, theoretical and practical, of the reinforcement of elastomers.[61–64]

Reinforcing agents

It is found that a general classification into strongly and weakly reinforcing can be obtained by plotting the extension of a filled elastomer against the so-called 'Mullins softening'—the reduction in elastic modulus caused by successive stretching. Appreciable softening, related to mechanical hysteresis, occurs only when there is true reinforcement.[65–67]

To compare fillers it is necessary that they shall be in the same matrix and cross-linked to approximately the same degree. Such a comparison is shown in Table 16 and is, at best, qualitative. It is seen that while the particle size or surface area of interaction between

TABLE 16
MULLINS SOFTENING FOR DIFFERENT MATERIALS[67]

Materials	Size	Softening
Calcium carbonate	50 nm	None
	5 000 nm	None
Clay	1 200 nm	Slight
Silica	20 nm	Strong
	100 nm	Medium
Silicate	30 nm	Strong
HAF carbon black	30 nm	Strong

filler and elastomer is the prime factor, in this particular matrix it is not the only one. It may be the bulk chemical nature of the filler which is involved or the nature of its surface or the degree of dispersion, or other factors. Careful work using carbon black of constant specific surface and surface chemical characteristics but differing in the degree of 'structure' (*see* below) shows that tensile strength and rupture energy are functions of the product of carbon black concentration and a structure-dependent factor which correlates with the oil absorption of the black.[68] Agglomeration and particle–particle interactions are thus involved, probably in addition to the influence of rubber occluded within the agglomerate. Reinforcement can arise only in these two ways, through interaction of filler particles with either the matrix or with other filler particles. It is, therefore, not surprising that strength and wear-resistance are often markedly affected by the degree of dispersion of the filler. Such a dispersion is often far from ideal and examination of a vulcanisate containing, say, 30 per cent of filler usually shows regions of much higher and lower filler content. Sometimes clouds of filler of intertwined coherent structure can be seen[69] and these lead to appreciable filler–filler interaction. Aggregates of filler of above about 10 μm seem to be most harmful; in some cases, tensile strength and wear-resistance increase approximately in proportion to the percentage of filler present in aggregates less than 9 μm in diameter.[70] To achieve such good dispersion requires rigorous control.[71] Examination of the fracture surface of an elastomer filled with carbon black shows the presence of a larger number of carbon particles than would be expected. Further, an unexpectedly high proportion of these particles are not bonded to the matrix.[72] It is possible that there is here a clue to the way in which filler particles reinforce. High stresses around the filler particles cause failure at, or near, the interface. These internal failures, which may cause dissipation of energy, thus increasing the work of tearing or failure, are subsequently linked by the course of the tear. If this is so, the need is shown for a good filler–elastomer bond so that failure would occur only at high stress. Such a bond must, at least in part, depend on the nature of the filler.

Carbon black as a reinforcement

Carbon black is by far the most utilised reinforcement in rubbers. Thousands of millions of pounds are used annually.[73] A very wide variety of blacks is available, classified by method of manufacture or so-called 'structure'. The available types and the progression of new types have been reviewed.[74,75] 'Structure' is a feature well known in carbon blacks and refers to the extent to which the primary particles are bound together in chains. It is quite distinct from reversible aggregation due to Van der Waals forces, etc. It is widely believed that the particles are fused together or share micro-crystallites as a result of collision at high temperatures while in a plastic state,[76] but that the necks between particles are broken during incorporation in the elastomer. Recent work has covered the effect of structure on the reinforcement action of carbon black.[62]

Carbon black interacts with the elastomer at two stages in the processing, when added to the raw rubber during milling and during the vulcanisation process. Addition of reinforcer to the raw rubber leads to the formation of 'bound rubber'—a gel containing both rubber and reinforcement and which is insoluble in the usual solvents for the elastomer. Bound rubber is usually pictured as hard islands (with carbon cores) set in a matrix of unchanged elastomer. The greater apparent degree of cross-linking within these islands includes a contribution from polymer-filler bonds formed during milling. The later increase in cross-linking during vulcanisation involves interaction of functional groups, surface active sites, etc., with each other and with additives under the influence of increased temperature. Recent views on the types of bonding represented in both these stages have been reviewed.[76] It is fairly certain that, with carbon black in an organic elastomer, bound rubber contains both chemical and physical bonding of particle to rubber, but the relative importance of these two forms is not yet settled and may vary with the particular fillers and elastomers in question, and also with the degree of stress or elongation. It is, however, worth noting that reinforcement with carbon black is remarkably insensitive to the chemistry of the elastomer involved. Physical bonding involves adsorption of the elastomer on to the surface of the carbon black and is, in itself, sufficient to produce strengthening. Since this type of adsorption is not localised, molecular realignment with slippage on the surface of the black may accommodate stress and dissipate energy as heat. Chemical interactions between the surface of the carbon black and the elastomer involve some of the wide variety of groupings found on the surface of the carbon black and are of three main types:

—radical trapping by acceptor sites on the carbon;

—radical initiation by hydrogen transfer to the polymer;

—ionic type bonding after proton transfer from surface acidic groups on the carbon black to points of unsaturation in the polymer.

Because of the wide variety of surface groupings on the carbon black and their intimate compatibility with the organic elastomer, it it possible that carbon black is, in some respects, unique and that such chemical interactions do not occur to a similar extent with other filler–elastomer combinations.

The relationship between the properties of a reinforced elastomer and the salient properties of the particular carbon black used for reinforcement is somewhat obscured by the influences of, among other factors,

—the nature of the elastomer;

—the degree of cure;

—the presence of 'promoters' for reinforcement;

—the selection of curing agents;

—possible interactions between additives;

—the degree of dispersion of reinforcers.

Interrelationships between various properties of a number of reinforced elastomers and between properties and reinforcement loadings for various compounding recipes abound in the commercial literature and some of these have been summarised.[77] A useful table compares the average reinforcement in tyre treads of a natural and one synthetic rubber for different carbon blacks when compounded to give greatest abrasion resistance.[77] The influence of particle size is paramount, but that of 'structure' can also be seen.

Carbon blacks are not much used in silicone rubbers and there is little available information on the degree of reinforcement which may be produced. What data there is shows no evidence of chemical bonding between filler and elastomer[78] and suggests that carbon black functions in a manner similar to that of white fillers in silicone rubbers.

Fine silica, silicates and other white reinforcements

Apart from carbon blacks, the only fillers classified as fully reinforcing are specially-prepared fine particle silicas and silicates of aluminium and calcium. Other oxides, such as alumina and zinc oxide, have been used experimentally. Such finely-divided fillers are usually prepared by precipitation or condensation routes. The total produced is less than one-tenth of that of carbon black, and about two-thirds of this is silica.[79]

The reinforcing action of non-black fillers is even less well characterised and understood than that of carbon blacks. There are many points of similarity with blacks (*e.g.* formation of chemically-bound rubber and modification of viscous properties), but also many differences.[79] White fillers are not capable of providing the wide range of surface groupings found on carbon black. On silica only silanol and siloxane groups are found. The surfaces of silica and silicates are more polar than those of carbon black and are often considerably affected by the presence of small amounts of adsorbed moisture. To date there appears to be no clear-cut evidence for chemical interaction between a white filler and an elastomer whether carbon-based or not, but it must be recognised that different mechanisms of reinforcement may predominate at different stresses or in different reinforcement–elastomer combinations of the same type.

Attempts have been made to compare different white fillers, both among themselves and with a high-area furnace carbon black when set in a series of organic elastomers. To try to make a valid comparison, for each elastomer the same basic recipe was used with all fillers, adjusting only the curing conditions when this was necessary. No attempt was made to achieve optimum properties for any combination.[80] Unless the comparison is made in this way, the complications of formulating and compounding and modifying the silica may produce quite different degrees of reinforcement from the same silica and elastomer.[81] The truly reinforcing white fillers produced increased hardness and tear-resistance and decreased resilience in all the elastomers examined. Tensile strength usually increased and, on occasion, passed through a maximum as the reinforcement loading was increased. As with carbon blacks, high surface area is required to give reinforcement and silica and alumina produced by vapour phase reaction were most effective. In general terms, the carbon black showed higher tensile strengths and moduli and lower elongations, but the ageing characteristics of white-filled appeared to be superior to those of black-filled compounds. Because of increased generation of heat and hardening during the blending of white fillers with elastomers, dispersion is more difficult and may be less efficient, in spite of the use of softening additives.

Little is known about any possible chemical interactions between white fillers and carbon elastomers. Reaction is reported to occur between an alumina surface and some compounds such as butadiene containing carbon–carbon double bonds. In this connection it is interesting to note that alumina is readily incorporated into poly-butadiene elastomers and produces an especially good balance of properties.[80]

White fillers find acceptance in elastomers for use in the pharmaceutical and food and paper industries and elsewhere where 'colour appeal' is important. The high electrical resistance of elastomers reinforced with white fillers, as compared with those containing carbon black, is often used to advantage in highly insulating rubbers. Rubber shoe heels and soles are usually reinforced with silica with a low proportion of carbon black as colorant. In this way 'scuff' marking is greatly reduced. Mixed black and white fillers are sometimes used in tyre treads to improve tear-resistance and crack-growth behaviour.[80]

In the reinforcement of silicone rubbers white fillers are very largely used. In this way translucency can be preserved and, with suitable dyes or pigments, colour can be controlled. Uses of silicone rubbers depend on good high- and low-temperature properties and high electrical resistance. Fine white fillers can lead to reinforcement and comparisons have been made between a number of different silicas (including some which had been esterified and were therefore hydrophobic) and one or two other white fillers in a few specific formulations.[80,81,82] The main findings are similar to those in other comparisons of reinforcers in not being completely clear cut, and may be summarised as follows:

—increased surface area of filler increases the elastic modulus and the resistance to tear propagation;

—hardness increases almost linearly with filler content;

—each filler (different surface area) shows optimum tensile strength at a different concentration of filler. For the hydrophobic silica a higher concentration is needed to produce maximum tensile strength, but, since the maximum is less sharp, more variation is permissible in the degree of blending;

—filler concentration influences hardness in a more striking way than does surface area;

—surface area/tensile strength curves flatten as the surface area increases.

It is difficult to obtain good dispersion of filler in the soft silicone rubber and considerable agglomeration occurs[83] (as it does with carbon black). In this respect a hydrophobic silica may be expected to show less tendency to agglomerate. In this work the range of matrices was exceedingly limited and no effort was made to control the relative proportions of silanol and siloxane groups on the surface of the silica.

The interaction between silicone rubber and white fillers has been very little studied. There is no evidence for permanent chemical linkages. It is thought that there is initially some attachment of the polymer chains to a silica filler, but that, subsequently, in time,[84] or when a small stress is applied,[78] the polysiloxane chains reorient to form a coating or cage around the silica particle so that reinforcement is then the result of interference by the filler with the free movements of the chains. When the filler particles are too large to be contained in the network, reinforcement is very much weaker. A somewhat similar mechanism has been proposed to account for the strengthening of an ethylene/propylene terpolymer by calcined clay used as a filler with hydrolysis of the silane keying agent producing silanols and polysiloxane.[85]

Future developments

There is much yet to be learned as to why, and how, fine particulate fillers influence the properties of various elastomers. More must be learnt concerning the nature of the filler–elastomer interactions under various conditions and how these interactions may be modified by altering the surface groupings on the filler. At present the influence of the nature of the filler—its strength or hardness, etc., is a completely unknown factor. When knowledge of this sort is available it will become more realistic to try to select a filler for a particular requirement. At the moment the almost universal presence of adsorbed water on the surface of the filler appears to be almost ignored, yet such water may play as vital a role in this field as it does in catalysis.

All this information has to be gained in systems which are highly advanced technologically and in which a very wide range of additives is used for various purposes. Simplification is required and in this connection the controlled use of irradiation to produce cross-linking must be welcomed.

It has been widely assumed that a uniform dispersion of filler is required, and, by using suitably-coated fillers, dispersion may be made easier, but it may be that some filler–filler interaction is desirable. It remains to be seen whether suitable after-treatment of crystalline fine powders prepared by a condensation process might result in the sharing of crystallite planes, and, if so, whether such material would prove to be a superior reinforcement in elastomers. These are some of the directions in which future development may be expected.

REINFORCEMENT OF THERMOPLASTICS AND THERMOSETTING PLASTICS

There has been relatively little detailed study of fine particulate reinforcement in either thermosetting or thermoplastic matrices. In thermosets the degree of cross-linking is usually high and such materials fail by brittle fracture. The most important reinforcers are fibrous, but particulate fillers are used for special purposes (*e.g.* alumina and silicon carbide impart abrasion resistance and graphite imparts electrical conductivity) or in order to cheapen the product, but, for these purposes, fine powders are not required.[86] Fine silica is added to control viscosity and impart thixotropy, but, as far as is known, there is no evidence that it exerts much reinforcing action.[87] On the other hand, fine titania has been shown to increase tensile strength in a cross-linked epoxy resin at least at certain relative humidities.[88] No attempt appears to have been made to ensure good wetting or other bonding of filler to matrix, and this area may well repay study.

The effect of the use of fillers in linear thermoplastics is much less than when the polymer is cross-linked. Thus, although a reinforcing carbon black, in common with coarser powders, increases the stiffness of polyethylene, this is accompanied by a decrease, rather than an increase, in tensile strength.[89] On the other hand, fine titania appears to somewhat increase tensile strength in polyvinyl acetate[90] while an attempt to correlate changes in the tensile strength of polyethylene containing different fine fillers with the particle size of the filler showed a slight increase in strength with the finest filler.[91] A surface treated kaolin which has been converted to an organophilic form has been shown to have little effect on the tensile strength of polypropylene, although some interesting optical effects were observed.[92]

Again, in all this work no special attempt was made to ensure wetting of the filler or any other type of bonding, and this could account for the somewhat different findings. Other factors may, however, be responsible, for example slight changes in the degree of crystallinity of the polymer, presence of plasticiser, etc. Whilst study of the bonding of filler to matrix is once again needed, such complicating factors as those mentioned above are likely materially to increase the difficulty of such work.

CERMETS AND CEMENTED CARBIDES

These types of metal–ceramic composite contain much higher proportions of ceramic phase than dispersion-strengthened metals.[93] Cermets normally contain 50–85 per cent of ceramic, whilst cemented ceramics contain more than 85 per cent ceramic, although the distinction is quite arbitrary. Industrially cemented carbides are the more important of these compounds with a cobalt–tungsten carbide or cobalt–mixed carbides of tungsten, titanium, tantalum, etc., predominating.[94] In terms of practical usefulness these materials represent an attempt to use the hardness of some ceramics whilst increasing somewhat the impact resistance by setting the ceramics in a metallic matrix. There are considerable differences in properties, depending on whether or not the metallic phase wets the ceramic.[95] In general, oxides are not wetted by metals (this, in itself, leads to the use of metal–oxide cermets as containers for molten metals) while carbides are so wetted by a number of metals.[96] Consequently, carbide–metal bonding is stronger than that between oxide and metal.

There are three interrelated important parameters in cemented carbides; hardness, rupture strength and toughness. At present, the main need is for increased toughness without detriment to the other two. The ideal microstructure for optimum strength is thought to be a uniform dispersion of uniform small ceramic particles separated by thin films of metallic binder.[97] In this way, virtually all the metal can be within the influence of the high modulus dispersed particle surfaces, plastic flow is restrained, and the metal cannot fail in the usual ductile manner. When the metal layer is somewhat thicker it will be less strong but more ductile, while, if very thick, pinning of dislocations will be more important than plastic restraint as in a dispersion-strengthened metal. When the metal phase is continuous, differential contraction after sintering will result in compressive stresses on the ceramic phase. The thickness of the metal layer is a function of the overall composition and the size of the ceramic particles. The rupture strength of tungsten carbide–cobalt reaches a maximum when the binder film has an average thickness of 0·3–0·6 μm.[98] The recent introduction of a cemented carbide containing

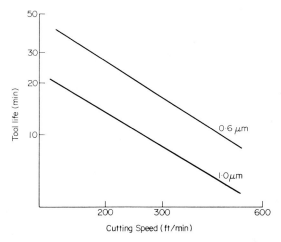

Fig. 23. Superior tool life of fine grained WC–Co alloys in machining cast iron.[103]

anisometric carbide grains, submicrometre in at least one dimension and triangular or platelet in shape as a result of hot working or, to some extent, directional freezing,[99] has led to improved relationships between the important properties.

Some solubility of the carbide in the molten metal matrix aids densification and improves bonding between the two phases. As a result of grain growth during the sintering or hot pressing process, the finest carbide particles now commercially available in cobalt are about 0·8 μm in diameter and considerable care must be taken to achieve this. The improvement in tool life obtainable by reducing the average grain size of the tungsten carbide from 1·0 μm to 0·6 μm[103] is shown in Fig. 23. Although outside the scope of this discussion, the stoichiometry of cemented carbides is quite critical since phases other than that desired (WC) may react at the interface with the metal matrix.

In applications in which the tool tip reaches high temperatures (700–800°C in high pressure, high speed, cutting of steels) diffusion occurs fairly readily through the metal matrix and there may be interchange of material between the cemented carbide tip and the workpiece. Titanium carbide has a low solubility in steel at these working temperatures and is usually incorporated, along with other carbides, into the hard tip. In an attempt to reduce brittleness, alloyed metal matrices are usually employed.[100] Tool tips of these types are superior to those of, for example, alumina, which are too brittle for many milling and machining operations.[101]

The use of submicrometre carbide powders as starting materials appears to show some advantages, for example, fracture then lies wholly within the metal phase or at the interface.[102] Hardness and abrasion resistance are improved with little loss of toughness.[103] Further advances along these lines are to be expected.

REFERENCES

1. Brown, L. M. and Ham, R. K. (1971). Dislocation-particle interactions. *Strengthening methods in crystals,* edited by A. Kelly and R. B. Nicholson, Applied Science Publishers, London, 12.
2. Nicholson, R. B. (1971). Strong microstructures from the solid state. *Strengthening methods in crystals,* edited by A. Kelly and R. B. Nicholson, Applied Science Publishers, London, 535.
3. Kelly, A. (1966). *Strong solids,* Oxford Univ. Press, 95.
4. Grant, N. J. (1966). Dispersion strengthening. *Strengthening mechanisms,* edited by J. J. Burke, N. L. Reed and V. Weiss, Syracuse Univ. Press, 63.
5. Ansell, G. S. (1965). Mechanical properties of two phase alloys. *Physical metallurgy,* edited by R. W. Cahn, Amsterdam, North Holland Publishing Co., 887.
6. Grant, N. J. (1964). Dispersed phase strengthening. *Strengthening of metals,* edited by D. Peckner, Reinhold, New York, 163.
7. Ansell, G. S. (1968). The mechanism of dispersion strengthening: a review. *Oxide dispersion strengthening,* edited by G. S. Ansell, T. D. Cooper and F. V. Lenel, Gordon and Breach, New York, 61.
8. Ashby, M. F. (1968). The theory of the critical shear stress and work hardening of dispersion-hardened crystals. *Oxide dispersion strengthening,* edited by G. S. Ansell, T. D. Cooper and F. V. Lenel, Gordon and Breach, New York, 143.
9. Ault, G. M. and Burte, H. M. (1968). Technical applications for oxide

dispersion-strengthened metals. *Oxide dispersion strengthening*, edited by G. S. Ansell, T. D. Cooper and F. V. Lenel, Gordon and Breach, New York, 3.

10. Sherby, O. D. and Burke, P. M. (1967). *Prog. Mater. Sci.*, **13**, (7), 325.
11. Fisher, J. C., Hart, E. W. and Pry, R. H. (1953). *Acta metall.*, **1**, 336.
12. Preston, O. and Grant, N. J. (1961). *Trans. Amer. Inst. Mining metall. Engrs metall. Soc.*, **221**, 164.
13. Hansen, N. (1969). *Acta metall.*, **17**, 637.
14. Ashby, M. F. and Smith, G. C. (1962/1963). *J. Inst. Metals*, **91**, 182.
15. Williams, D. M. and Smith, G. C. (1968). A study of oxide particles and oxide–matrix interface in copper. *Oxide dispersion strengthening*, edited by G. S. Ansell, T. D. Cooper and F. V. Lenel, Gordon and Breach, New York, 509.
16. Palmer, I. G. and Smith, G. C. (1968). Fracture of internally oxidised copper-alloys. *Oxide dispersion strengthening*, edited by G. S. Ansell, T. D. Cooper and F. V. Lenel, Gordon and Breach, New York, 253.
17. Edelson, B. E. and Baldwin, W. M. (1962). *Trans. Amer. Soc. Metals*, **55**, 230.
18. Seybolt, A. U. (1968). Some observations on the stability of oxide dispersions in metals. *Oxide dispersion strengthening*, edited by G. S. Ansell, T. D. Cooper and F. V. Lenel, Gordon and Breach, New York, 469.
19. Worn, D. K. and Marton, S. F. (1961). Some properties of nickel containing a dispersed phase of thoria. *Powder metallurgy*, edited by W. Leszinski, Interscience, New York, 309.
20. Komatsu, N. and Grant, N. J. (1962). *Trans. Amer. Inst. Mining metall. Engrs metall. Soc.*, **224**, 705.
21. Walter, J. L. and Seybolt, A. U. (1969). *Trans. Amer. Inst. Mining metall. Engrs metall. Soc.*, **245**, 1093.
22. Olsen, R. J., Judd, G. and Ansell, G. S. (1971). *Metallurgical Trans.*, **2**, 1353.
23. Bufferd, A. S. (1966). Preparation of metal–oxide dispersion composites, *Strengthening mechanisms—metals and ceramics*, edited by J. J. Burke, N. L. Reed and V. Weiss, Syracuse Univ. Press, 531.
24. Bufferd, A. S. (1968). *Fibre Sci. Techn.*, **1**, 63.
25. Ashall, D. W. and Evans, P. E. (1967). *Powder Metall.*, **10**, (20), 326.
26. Gilbert, A., Ratcliff, J. L. and Warke, W. R. (1965). *Trans. Amer. Soc. Metals*, **58**, 142.
27. Huet, J. J. (1967). *Powder Met.*, (20), 208.
28. Weeton, J. W. and Quatinetz, M. (1968). Cleaning and stabilisation of dispersion-strengthened materials. *Oxide dispersion strengthening* edited by G. S. Ansell, T. D. Cooper and F. V. Lenel, New York, Gordon and Breach, 751.
29. Benjamin, J. S. (1970). *Metallurgical Trans.*, **1**, 2943.
30. Arnold, D. B. and Klingler, L. J. (1968). Dispersion-strengthened nickel base alloys. *Oxide dispersion strengthening* edited by G. S. Ansell, T. D. Cooper and F. V. Lenel, Gordon and Breach, New York, 611.
31. Cheney, R. F. and Smith, J. S. (1968). Oxide-strengthened alloys by the selective reduction of spray-dried mixtures. *Oxide dispersion strengthening* edited by G. S. Ansell, T. D. Cooper and F. V. Lenel, Gordon and Breach, New York, 637.
32. Murphy, R. and Grant, N. J. (1962). *Powder Metall.* (10), 1.
33. Chin, L. L. J. and Grant, N. J. (1967). *Powder Metall.*, **10**, (20), 344.
34. Gimpl, M. L. and Fuschillo, N. (1968). Dispersion-hardened alloys made

by vapour plating and chemical precipitation techniques. *Oxide dispersion strengthening,* edited by G. S. Ansell, T. D. Cooper and F. V. Lenel, Gordon and Breach, New York, 719.

35. Dunkerley, F. J., Leavenworth, H. W. and Eichelman, G. E. (1968). Electrodeposition of dispersion-strengthened alloys. *Oxide dispersion strengthening* edited by G. S. Ansell, T. D. Cooper and F. V. Lenel, Gordon and Breach, New York, 695.
36. Bloch, E. A. (1961). *Metall. Revs,* **6**, 193.
37. Evans, D. J. I., Huffman, H. R. and Warner, J. P. (1964). *Mater. Des. Engg,* **59**, 105.
38. Hirschhorn, J. S. and Lenel, F. V. (1966). *Trans. Amer. Soc. Metals,* **59**, 208.
39. Murphy, R. J. and Grant, N. J. (1967). *Trans. Amer. Soc. Metals,* **60**, 29.
40. Mukherjee, A. K. and Martin, J. W. (1960). *J. less common Metals,* **2**, 392; (1961), **3**, 216.
41. Kindlimann, L. E. and Ansell, G. S. (1970). *Metallurgical Trans.,* **1**, 507.
42. Gatti, A. (1959). *Trans. Amer. Inst. Mining metall. Engrs metall. Soc.,* **215**, 753.
43. Ghodsi, M. (1970). *Metallurgie,* **10**, 105.
44. Beghi, G. (1967). *J. nucl. Mater.,* **23**, 241.
45. Seeman, H. J. and Staats, H. (1968). *Zeit. Metallk.,* **59**, 347.
46. Giggins, C. S. and Pettit, F. S. (1971). *Metallurgical Trans.,* **2**, 1071.
47. Clegg, M. A. and Lund, J. A. (1971). *Metallurgical Trans.,* **2**, 2495.
48. Hahn, G. T. and Rosenfield, A. R. (1967). *Trans. Amer. Inst. Mining metall. Engrs metall. Soc.,* **239**, 669.
49. Hahn, G. T. and Rosenfield, A. R. (1966). *Acta metall.,* **14**, 1815.
50. Ryan, N. E. and Johnstone, S. T. M. (1965). *J. Less Common Metals,* **8**, 159.
51. McHugh, C. O., Whalen, T. J. and Humenik, M. (1966). *J. Amer. ceram. Soc.,* **49**, 486.
52. Anon. (1968). *Steel,* **162**, (13), 11.
53. Rankin, D. T., Stiglich, J. J., Petrak, D. R. and Ruh, R. (1971). *J. Amer. ceram. Soc.,* **54**, 277.
54. Coble, R. L. and Burke, J. E. (1963). Sintering in ceramics, *Prog. ceram. Sci.,* **3**, edited by J. E. Burke, Pergamon, Oxford, 197.
55. Ervin, G. (1971). *Bull. Amer. ceram. Soc.,* **50**, 659; (1971) *J. Amer. ceram. Soc.,* **54**, 46.
56. Payne, A. R. (1966). Elastomer systems. *Composite materials,* edited by L. Holliday, Elsevier, Amsterdam, 290.
57. Gehman, S. D. (1965). Mechanism of tearing and abrasion in reinforced elastomers. *Reinforcement of elastomers,* edited by G. Kraus, Interscience, New York, 23.
58. Payne, A. R. and Whittaker, R. E. (1971). *J. appl. Poly. Sci.,* **15**, 1941.
59. Harwood, J. A. C. and Payne, A. R. (1968). *J. appl. Poly. Sci.,* **12**, 889.
60. Weissert, F. C. (1969). *Ind. Eng. Chem.,* **61**, (8), 53.
61. Payne, A. R. and Whittaker, R. E. (1970). *Composites,* **1**, 203.
62. Blanchard, A. F. (1971). *Rubber J.,* (2), 44; (3), 25.
63. Bulgin, D. (1971). *Composites,* **2**, 165.
64. Kraus, G. (1971). *Adv. Polymer Sci.,* **8**, 155.
65. Bueche, F. (1960). *J. appl. Polym. Sci.,* **4**, 107.
66. Mullins, L. and Tobin, N. R. (1957). *Rubber Chem. Technol.,* **30**, 555.
67. Sellars, J. W. and Toonder, F. E. (1965). Reinforcing fine particle silicas and silicates. *Reinforcement of elastomers,* edited by G. Kraus, Interscience,

New York, 405. Bartrug, N. G., Fear, R. H. and Wagner, M. P. (1964). *Rubber Age*, **96**, 405.
68. Kraus, G. (1971). *J. appl. Polymer Sci.*, **15**, 1679.
69. Hess, W. M. and Ford, F. P. (1963). *Rubber Chem. Technol.*, **36**, 1175.
70. Sweitzer, C. W. (1962). *Rubber Plast. Weekly*, **142**, (8), 283.
71. Boonstra, B. B. and Medalia, A. I. (1963). *Rubber Age*, **46**, 892; **47**, 82.
72. Andrews, E. H. and Walsh, A. (1958). *Proc. phys. Soc.*, **72**, 42.
73. Smith, W. R. and Bean, D. C. (1964). Carbon black. *Encyclopedia of chemical technology*, second edition, **4**, edited by H. F. Mark, J. J. McKetta and D. F. Othmer, Interscience, New York, 243.
74. Honak, E. R. (1969). *Kunstst.-Rundschau*, **16**, 1.
75. Dannenberg, E. M. (1971). *J. Inst. Rubber Ind.*, **5**, 190.
76. Deviney, M. L. (1969). *Adv. Colloid Interf. Sci.*, **2**, 237: Kraus, G. (1965). *Rubber Chem. Technol.*, **38**, 1070.
77. Studebaker, M. L. (1965). Compounding with carbon black. *Reinforcement of elastomers*, edited by G. Kraus, Interscience, New York, 319; Studebaker, M. L. (1957). *Rubber Chem. Technol.*, **30**, 1400.
78. Baker, D., Charlesby, A. and Morris, J. (1968). *Polymer*, **9**, 437.
79. Smith, D. A. (1968). *Rubber J.*, **150**, (9), 54.
80. Westlinning, H. and Fleischhauer, H. (1965). Properties of white reinforcing fillers in elastomers. *Reinforcement of elastomers*, edited by G. Kraus, Interscience, New York, 425.
81. Degussa Technical Service Lab., (1968). *Rubber Wld.*, **158**, (4), 59.
82. Bode, R., Ferch, H. and Fratzscher, H. (1967). *Kaut. Gummi Kunstst.*, **20**, 699.
83. Galanti, A. V. and Sperling, L. H. (1970). *J. appl. Polymer Sci.*, **14**, 2785.
84. Southwart, D. W. and Hunt, T. (1968). *J. Inst. Rubber Ind.*, **2**, 77, 79, 140.
85. Schwaber, D. M. and Rodriguez, F. (1967). *Rubber Plast. Age*, **48**, 1081.
86. Loewenstein, K. L. (1966). Glassy systems. *Composite materials*, edited by L. Holliday, Elsevier, Amsterdam, 129.
87. Schue, G. K. (1969). *S.P.E. Jl.*, **25**, (7), 40.
88. Galperin, I., Arnheim, W. and Kwei, T. K. (1965). *J. appl. Polym. Sci.*, **9**, 3215.
89. Boonstra, B. B. (1965). Reinforcement of polyethylene. *Reinforcement of elastomers*, edited by G. A. Kraus, Interscience, New York, 529.
90. Galperin, I. and Kwei, T. K. (1966). *J. appl. Polym. Sci.*, **10**, 681. Galperin, I. and Arnheim, W. (1967). *J. appl. Polym. Sci.*, **11**, 1259.
91. Alter, H. (1965). *J. appl. Polym. Sci.*, **9**, 1525.
92. Freeport Mineral Co., (1971). *SPE Jl.*, **27**, (9), 16.
93. Schwartzkopf, P. and Kieffer, R. (1960). *Cemented carbides*, Macmillan, New York. (1960). *Cermets*, edited by J. R. Tinklepaugh and W. B. Crandall, Reinhold, New York.
94. Morral, F. R. (1966). Refractory compounds as structural materials— Cobalt-bonded refractory compounds. *Fundamentals of refractory compounds*, edited by H. H. Hausner and M. G. Bowman, Plenum Press, New York, 229.
95. Petzow, G., Claussen, N. and Exner, H. E. (1968). *Zeit. Metallk.*, **59**, 170.
96. Ramqvist, L. (1965). *Int. J. Powder Metall.*, **1**, (4), 2.
97. Stoops, R. F. (1965). *N. Carolina State College, Dept. Engg Res., Bull.* 82.
98. Gurland, J. (1963). *Trans. Amer. Inst. Mining metall. Engrs metall. Soc.*, **227**, 1146.
99. Meadows, G. W. U.S. patent 3,515,540 (June 2, 1970); 3,525,610 (August 25, 1970).

100. Todd, A. G. and Stafford, J. S. (1967). *AEI Engg Rev.,* **7,** 154.
101. Brewer, R. C. (1957). *Engg Dig.,* **18,** 381.
102. Hara, A., Nishikawa, T. and Yazu, S. (1970). *Planseeber. f. Pulvermet.,* **18,** 28.
103. Improved cemented carbides reduce wear and corrosion. (1968). *Mater. Engg,* **68,** (2), 30.

CHAPTER 6

FINE POWDERS IN SURFACE COATINGS

PAINTS, INKS, ETC.

Fine powders are used as pigments in surface coatings and make definite contributions to the properties of such coatings as paints, enamels and inks. Thus the chemical properties of the pigment assist in establishing the suitability of paints for external use, whilst physical properties have their place in determining sheen, texture, consistency and thickness of the coating. The field is very large, with its own textbooks (for example, references 1–3) and society journals from which further information may be sought. We are here concerned with the part played in coatings by the fine-particle powders and with the specific properties of these powders which make them suitable for use in such coatings.

A paint film is typically about 20 μm thick, so all pigment particles must be smaller than this, but the actual size may vary quite widely from perhaps 10 μm down to perhaps 0·01 μm. In letterpress or lithographic printing the ink film is less than 3 μm thick so the upper limit on particle size is more restrictive.[4] This upper limit applies not to the size of the ultimate particle, but to any agglomerate which may resist the forces of application and persist in the final coating. Thus, although many pigments used in surface coatings are not less than 1 μm in diameter, and therefore are not strictly within the scope of our consideration, it is true to say that pigments represent one of the two largest uses for submicrometre powders. A brief analysis of the influence of the pigment on the properties of a surface coating will show which are the important properties of the pigment.

Surface finish—sheen and texture of film

A high gloss exists when a beam of light falling on a surface is reflected directionally. A mirror may reflect directionally more than

107

90 per cent of incident light at high angles of incidence. Surface coatings do not usually reach this figure and it is typical of this type of reflection that, as the angle of incidence decreases from 90°, the amount of light reflected falls off. Gloss is imparted by a surface smooth over an extent much greater than the wavelength of light. In a surface coating it is largely a property of the air-binder or air-medium interface with the pigment only intruding when the particles cause irregularity in the reflecting surface. The pigment particles must therefore be small and must not be present in too high a concentration.

A matt surface on the other hand is, on the microscale, rough, and multiple reflection of an incident beam occurs, producing a much more even distribution of reflected light. This is usually the result of a much higher pigment concentration in the medium. Large particles produce uneven effects, but particles can be somewhat larger than those in a gloss finish.

Optical properties of the film

There are two mechanisms of the interaction of light with matter—absorption which converts incident radiation into heat, and scattering in which incident radiation is absorbed and re-emitted. Except for white material which shows only very slight absorption at any visible wavelength, both these forms of light attenuation occur simultaneously. The eye perceives colour in a surface coating as a result of the selective reflection or scattering of light incident upon the object. Under other circumstances light may be seen principally after transmission, and transmission also participates, rather indirectly, in the perception of colour in a surface film. White light incident on a pigment particle in such a film is partly scattered, partly transmitted and partly absorbed. The scattered and transmitted light may then be incident upon another pigment particle. Eventually, all unabsorbed radiation will emerge from the surface film. Any such radiation, which has been reflected by the substrate, will show the optical character of the substrate and will represent failure of the surface coating to obliterate or hide the substrate. This leads to the technological parameter 'hiding power' which is used for the comparison of films. A quantitative analysis somewhat on the above lines has led to the so-called Kubelka-Munk relationship:[5]

$$\frac{K_\lambda}{S_\lambda} = \frac{(1 - R_{\infty\lambda})^2}{2R_{\infty\lambda}}$$

where:
    K = the absorption coefficient;
    S = the scattering coefficient;
    R_{∞} = the reflectance of a film of such thickness that further increase has no effect;
    λ = a subscript that indicates that the quantities are dependent on wavelength.

Similar types of relationship have been found for a mixture of pigments, using a somewhat different approach.[6]

The scattering of light is dependent on the relationship between the refractive indices of pigment and binder. This is expressed by the Lorentz–Lorenz relation:

$$M = \frac{(n_P/n_B)^2 - 1}{(n_P/n_B)^2 + 2}$$

where M is the scattering and n_P and n_B are, respectively, the refractive indices of pigment and binder. For a white pigment in a colourless binder both n_P and n_B show slight variations with wavelength, so M is somewhat dependent on wavelength.

White pigmentation depends entirely on scatter, so the highest possible value of n_P is required if n_B is in the usual range for organic oil media. This is currently met by the use of the rutile form of titania ($n_P \sim 2 \cdot 76$) as pigment. In emulsion paints and distempers the scattering interface is predominantly between pigment and air and acceptable scatter can then be produced with lower values of n_P provided by whiting, silicates, etc. In considering the influence of particle size on scatter, it is necessary to distinguish two regions of particle size. When the particle diameter is appreciably greater than λ, the wavelength of light, the scattered intensity is approximately inversely proportional to d, the particle diameter. When d is appreciably less than the wavelength of light (*i.e.* in the region of Rayleigh scattering) total scattered light is given by:

$$S \propto \frac{d^6}{\lambda^4}$$

There is thus a particle size at which scatter is a maximum. This size is a complex function of both the relative index of refraction and the wavelength of light in the medium in which the particle is situated.

There have been a number of calculations of optimum size, some purely theoretical, others with experimental basis (*see* reference 7), and the results differ fairly widely (for titania in a typical paint medium from 0·2 μm to 0·4 μm). The decrease in scatter at sizes smaller than the optimum is quite sharp and since for a particular particle size in this region there is greater scatter of blue light than of longer wavelengths, white powders of smaller than optimum size show a bluish tinge. Presumably, if the particle size were continuously reduced, a point would eventually be reached when the particles would appear transparent since scatter would be so low. Thus, from a practical point of view, to achieve maximum scatter and a white colour quite a narrow range of particle size is required.

In all these considerations of scattering the dispersion of the pigment is assumed to be ideal. In practice this is not the case (*see* page 112).

Where coloured pigmentation is required, in order to obliterate the substrate a white pigment is often added to the coloured pigment to provide the necessary scatter. This dilution of the colouring pigment affects the final colour. The effect of variations of particle size on coloured pigmentation is complicated and is most often considered in an *ad hoc* manner by means of comparative 'tinting strength' tests. Both scattering and absorption are involved in such tests and it is only for carbon black that these two components have been separated;[8] for carbon black scattering falls off sharply below a maximum at about 0·5 μm while absorption reaches a maximum at about 0·25 μm. In general terms, it is necessary to distinguish between strong absorbers for which the absorption coefficient K is greater than $1/\lambda$, and weak aborbers where $K < (1/\lambda)$. Strong absorbers of particle size about 1 μm usually show a slight increase in apparent absorption with decreasing particle size, whilst weakly absorbing materials show greatly decreased absorption under the same conditions.[9] This difference in behaviour is explained by considering the degree of regular or surface reflection at the surface of the pigment particle and the depth of penetration and attenuation of radiation entering the pigment particle. Surface reflection increases with the coefficient of absorption according to the Fresnel equation which, for perpendicular incidence is:

$$\text{reflection} = \frac{[(n_\text{P}/n_\text{B}) - 1]^2 + K^2}{[(n_\text{P}/n_\text{B}) + 1]^2 + K^2}$$

Thus, for a very strongly absorbing material, a high proportion of the incident radiation is reflected at the surface and the depth of penetration into the particle is smaller by several orders of magnitude than for a weakly absorbing material. Reduction in the particle size, and thus in the depth of material which is available for the attenuation of radiation, is therefore likely to considerably reduce absorption in a weakly absorbing material, but to have very little effect in a strong absorbent.[10] The slight increase in absorption in the latter type of material is due to a small reduction in the proportion of light reflected at the surface—an effect which is negligible in the weak absorber. However, when the particles are not of uniform size and are viewed in poly- rather than mono-chromatic light, the theoretical relationships tend to lose their value. But, on a purely experimental basis, maxima in tinting strength have been found for prussian blue at 0·3 μm and for molybdate orange at 0·5 μm.[11] At wavelengths at which absorption occurs, the refractive index undergoes very marked increase.[12] What scatter there is with coloured pigments is thus strongly dependent on wavelength. The brilliance of a colour is dependent on the width of the wavelength band over which absorption occurs.[13] To give a pure primary colour, a narrow absorption band is required with a minimum of absorption outside that band. Other colours always show broad or multiple bands.

Pigment–medium interactions

A 'dry' pigment in the form of a powder usually consists of agglomerates of primary particles with air between and with a water layer coating the particles and perhaps condensed in the saddles between particles. The forces causing agglomeration are listed in Chapter 8. In addition, materials prepared by thermal decomposition or precipitation consist of aggregates of primary particles held together by solid bridging and requiring grinding before adding to a liquid medium. To avoid the need for this grinding, some precipitated pigments are now spray-dried to give a less coarse powder.

Apart from the danger that agglomerates will affect the smoothness of the surface coating, the extent of agglomeration affects the settling properties and the optical behaviour of the coating. Large agglomerates lead to rapid settling and poor optical performance in that the number of effective scattering centres falls. If agglomerates are broken down, the individual pigment particles may remain completely deflocculated after wetting, or they may come together in the medium to form flocs similar to the original agglomerates but with space originally filled with air now containing the liquid medium. The deflocculated pigment shows optimum optical properties with excellent levelling (obliteration of brush marks), but is inferior to the flocculated form in its tendency to sag and run (especially on vertical surfaces) and in the difficulty of application (for example by brush). Thus, the ideal is a compromise between flocculated and deflocculated, as shown in Fig. 24, but with initial complete wetting by the medium.[14]

The importance of the whole process of converting the dry pigment into a suitable form for application in a surface coating has recently been emphasised in a cogent review[15] which attempts to distill basic scientific principles from the hotch-potch of technological practice. There are three stages; wetting, deagglomeration and stabilisation. In wetting, the air or water–pigment interface is replaced by a medium–pigment interface. The standard theory of wetting is widely applicable. However, it must be emphasised that the nature of the pigment surface is, at best, ill-defined.

Such a surface will be energetically heterogeneous and may carry various atomic groupings. Unselective application of simple wetting theory can, therefore, lead to anomalies. In general, solids with a high surface free energy are wetted by liquids with lower surface free energies,[16] but the effect of surface adsorption on the surface free energy can vary considerably. The presence of pores in the pigment particles or of fine spaces between the particles will reduce the speed with which wetting occurs. Various combinations of medium, surfactant and other additive are usually employed in an attempt to produce a uniform stable dispersion, and the effect of these additives on wetting may be considerable.

Once the surface of the pigment is wetted, there is no great difficulty in breaking down aggregates or agglomerates by milling, grinding, etc., but, unless the well-dispersed system thus produced can be stabilised, it has little value. The forces leading to reagglomeration are essentially those causing agglomeration of the initial powder. Mechanisms for opposing these forces are based on either surface charge or an adsorbed layer. The surface charge concept involves the adsorption of counter ions on the surface of the pigment particle to 'neutralise' the zeta potential of the particle. The charged layer

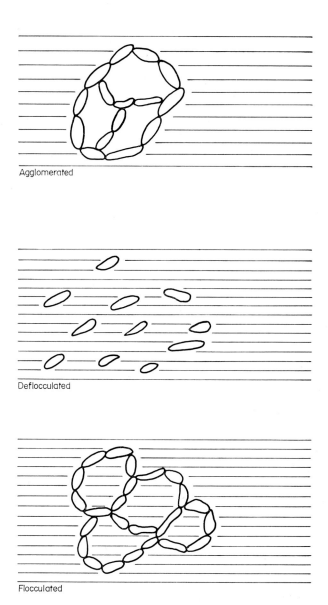

Agglomerated

Deflocculated

Flocculated

Fig. 24. *Properties of idealised dispersions of pigments in a medium.*[14]

gives rise to an electrical potential surrounding each particle, and the potentials of neighbouring particles repel each other. However, small particles have only small zeta potentials,[17] so stabilisation by this mechanism falls off rapidly when the particles are less than 1 μm in diameter. Under these circumstances the adsorption of suitable molecules on the pigment surface may increase stability either by reducing the agglomerating forces or by preventing the pigment particles from approaching each other by steric hindrance.[18] Although considerable effort has recently been devoted to studying the nature of these adsorbed layers, the technology of the production of suitable stable dispersions is very much on an *ad hoc* basis (*see* references 19 and 20). Recent work has shown that pigment flocculation may occur during the application and/or drying of the coating,[21] so this may open up a new approach to the problem.

Subsequent to the drying of the coating film there may be chemical reaction at the pigment–medium interface. Ultraviolet radiation has sufficient energy to break covalent bonds. Titania, the most common white pigment, absorbs in the ultraviolet and, in the presence of moisture, participates in what is, in effect, a photochemical reaction leading to oxidation of the organic medium at the interface.[22] The role of the titania may be either that of a direct supplier of activated oxygen atoms or of an activating agent enabling oxygen from the atmosphere to enter the reaction cycle.[23] Since the oxidation products are to some extent soluble, the organic binder is gradually washed away and the pigment particles left unbonded on the surface. This is the phenomenon of 'chalking'. Other deleterious results of these photochemical reactions may be fading of organic dyes incorporated with titania, or, in white coatings, a yellowing due to oxidation of the colourless organic binder. Because of reactions of these types virtually all titania pigments are surface-treated with silica or some other material such as alumina which itself absorbs ultraviolet radiation and harmlessly dissipates the energy.[22] The patent literature on the methods of applying such coatings and their properties is quite extensive.[24–29]

The performance of a paint film (its weatherability, fastness and stability) is largely governed by pigment–medium interaction to improve which pigments such as titania are constantly being modified. There is probably more to be understood concerning the pigment–medium interface than any other single sector of the pigment field.[30] Consequently, increasing use of modern techniques such as electron microscopy is likely to be especially fruitful.[31]

Control of rheology in surface coatings

Control of the rheology of surface coatings is very necessary. Thus a pigment would be useless if it gave sufficient opacity only in concentrations such that the coating was too stiff for application. This situation may be avoided by using a pigment with a high hiding power or high tinting strength. In the reverse situation, when the coating is too thin for easy application, addition of an extender will add 'body'. Extenders were cheap materials of lower refractive index than pigments, but recently very fine powders (especially silica) have come to be used to increase the stiffness of a coating. Such fine powders also show a degree of thixotropy when used in suitable organic media and this is desirable in both paints[32] and inks.[33]

Commercial pigments for surface coatings

Such pigments may be inorganic or organic. The more stable colours now available in the latter are leading to increased use of organic coloured pigments, as compared with inorganic. However, since the latter are often cheaper, such materials as the different forms of iron oxide are likely to continue to be used. In white pigments inorganics predominate. Titania is used in very large and increasing tonnage while zinc oxide, antimony oxide and lead-sulphate and barium sulphate-based pigments now find less use. For black coloration some form of carbon black is usually used.

As regards particle size, zinc and antimony oxides, which are usually produced by burning the corresponding elements, and titania, together with carbon black, are usually produced in sub-micrometre particle size. An extra step is necessary to increase the size of titania particles produced by the new process of oxidation of the chloride to the optimum for light scattering and to avoid the bluish tint. Synthetic iron oxides usually produced by calcination of ferrous sulphate can take on a wide range of colours from pink to dark bluish black. Differences in colour intensity and also slight differences in shade for the α-ferric oxide pigment can be correlated with particle size.[34]

PIGMENTS IN PLASTICS AND RUBBER

Pigments are added to these materials to mask unattractive natural colours, to appeal to the eye and sometimes for colour coding. Although the general principles are similar to those for surface coatings, the pigment can now be dispersed through a much greater thickness. Organic dyes are much used, but for thermosetting plastics inorganic pigments predominate, because at the temperatures needed for curing (up to 300°C) organic pigments are not stable.[35] White reinforcing pigments much used are titania, silica and zinc oxide, while carbon black is used more often as a reinforcement than a pigment.

COATING OF PAPER

Although the coating of paper has some similarity, in principle, with such surface coatings as paints, inks, etc., considered above (especially with regard to colouring), much paper is used as a surface for printing and as a recipient of ink coatings and other special coatings. Pigments are introduced into paper at two stages. The first is in the production of sheet paper where the pigment acts more as a filler, and the second when a true coating is applied to the sheet. Much of the pigmentation of paper and its opacity can be produced by the same submicrometre size materials added at either stage. We are here concerned with coating.

Colouring or masking action

The colouring action of a coating on paper is very similar to that in a paint which contains a high concentration of pigment. The main scattering interfaces are air–pigment and the coating is porous. The main particulate colorants are carbon black and white pigments such as titania, zinc sulphide, calcium sulphate and calcium carbonate with organic dyes as required. Again, titania is often used with a silicate coating which has a high absorption in the ultraviolet and thus helps to prevent yellowing.[36] Much of the field of pigmentation and coating has been covered by recent monographs.[37,38] The coating is normally quite porous as applied, and ink dries by soaking into the paper. The degree of porosity influences the viscosity requirement for any ink applied and both these properties are involved in the concept of the drying time of applied inks. In large-scale printing a lengthy drying time can be a great embarrassment.

Surface gloss

Both opacity and gloss are greatly improved by the use of clay as a filler/coater. When clay particles are less than about 2 μm in diameter they occur as hexagonal platelets. Such particles deposit with the flat of the platelets on the paper surface and this produces a relatively impervious, glossy and non-absorptive surface film.[38] This type of surface tends to produce more brilliant printing colours. Fine powders, lacking the anisotropic morphology of clays, produce a more matt finish on the paper, but allow fairly close control to be exercised over the rheology of the coating. This is particularly important when the coating is applied by roller when there is a high degree of shear between coating and roller. A thixotropic coating mix produces less surface roughness.

Special surface coatings

The use of finely-divided submicrometre silica as a base for other special coatings is exemplified in light-sensitive papers of the diazo type.[39] The silica forms a highly absorbent coating and the light sensitive diazo compound is fixed thereon instead of penetrating into the paper. This leads to improved light sensitivity, and improved definition of the image.

One of the most rapidly progressing techniques for office copying, a direct electrostatic method, is currently based on the use of zinc oxide powder as a photoconductor.[40] The principle of this type of xerographic process has been outlined.[41] The zinc oxide is dispersed in an insulating binder matrix. For the conductivity of the layer under illumination to reach a maximum it is necessary that the high dark resistivity is realised by low free charge-carrier concentration and not by low mobility of charge-carriers. This means that compensation of bulk impurities is required and, with fine zinc oxide, this can occur by adsorbed surface impurities, thus overcoming the difficulty of uniformly distributing an insoluble dopant. Dye sensitisation is used to broaden the spectral response. A study has been made[42] of methods for the burning of metallic zinc in order to produce oxide with suitably reproducible electrophotographic properties. All the oxides examined were of submicrometre size and exceedingly close monitoring of the process was needed to control purity, stoichiometry and, to some extent, particle size.

A further use of powders partly of submicrometre size is suggested by the fact that a coating of boron nitride in the particle size range 0·01 μm to 44 μm increases the thermal conductivity of the paper for use as a dielectric. The power handling capability of the component so produced is augmented.[43]

Possible future developments in fine powders used as surface coatings

With few exceptions, coloured pigment particles now in use are not fine powders as defined here. Apart from the cheap carbon black and iron oxide, most coloured pigments used in paints and inks are organic. This is due not to the inherent cheapness of organic pigments, but to the intensity of the colours which may be obtained and to the recently-developed light fastness. In plastics, inorganic coloured pigments will probably continue to be needed for thermal stability. Here the increasing cost of some (*e.g.* cadmium sulphoselenide) and the poor reproducibility of others (*e.g.* nickel titanate) may be expected to act as spurs in the search for new inorganic pigments. Whilst it is possible to predict, with some degree of assurance, the general colour of an organic compound of a given structure, the connection between colour and structure for an inorganic compound is frequently much more complicated. Explanation of an observed colour is usually possible, but prediction is rarely possible at the present time. Future developments will, in the long run, change this state of affairs.

At the moment, the particle size of coloured pigments is not usually closely controlled. In future it will probably become possible to separate the scattering and absorption functions with respect to particle size, and to optimise one or other function by controlling particle size.

White pigments are inorganic. The high refractive index of titania confers increased scattering power on the interface between titania and the medium. However, titania needs to be coated in order to reduce photochemical reaction with the organic medium, leading to discoloration or chalking. A non-chalking white pigment of equivalent scattering power would offer advantages. Silicon carbide has a similar refractive index, and can occur as plate-shaped particles which would give good abrasion resistance to the coating. However, silicon carbide is a semi-conductor and colour control may depend critically on the concentration of impurities so that a reproducible white colour may not be obtainable. Further, it is doubtful whether silicon carbide can show sufficient surface activity to interact to a sufficient extent with the medium.

Much work has already been done on the dispersion of pigment particles in a medium. Very recently, electron microscopy has been used to examine the dispersion of pigment particles in a paint film, and attempts will probably be made to correlate this final dispersion with that in the coating as applied and thus with the surface chemistry of the pigment particles themselves. Developments such as these may be expected to put on to a firmer scientific base the pigment field, which is already technologically well advanced.

REFERENCES

1. *Technology of paints, varnishes and lacquers* (1968). Edited by C. R. Martens, Reinhold, New York.
2. *Pigments—an introduction to their physical chemistry* (1967). Edited by D. Patterson, Applied Science Publishers, London.
3. Nylen, P. and Sunderland, E. (1965). *Modern surface coatings,* Interscience, New York.
4. Bowles, R. F. (1961). Pigments and extenders for use in inks. *Powders in industry,* London, Society for Chemical Industry, 185.
5. Kubelka, P. and Munk, F. (1931). *Zeit. tech. Phys.,* **12**, 593.
6. Duncan, D. R. (1940). *Proc. phys. Soc.,* **52**, 390.
7. Mitton, P. B., Vejnoska, L. W. and Frederick, M. (1961). *Off. Dig. Fed. Socs. Paint Technol,* **33**, 1264; (1962) **34**, 73.
8. Hodkinson, J. R. (1964). *J. opt. Soc. Amer.,* **54**, 846.
   Hodkinson, J. R. (1966). The optical measurement of aerosols. *Aerosol science,* edited by C. N. Davies, Academic Press, New York, 287.
9. Wendlandt, W. W. and Hecht, H. G. (1966). *Reflectance spectroscopy,* Interscience, New York, 51.
10. Kortum, G. and Vogel, J. (1958). *Zeit. phys. Chem. Frankfurt,* **18**, 230.
11. Barrick, G. (1941). *Amer. Paint J. Convention Daily* (October) 12.
12. Patterson, D. The colour of pigment crystals, reference 2, 55.
13. Feitknecht, W. The theory of colour in inorganic substance, reference 2, 3.
14. Ensminger, R. I. (1963). *Off. Dig. Fed. Socs. Paint Technol.,* **35**, 71.
15. Buttignol, V. and Gerhart, H. L. (1968). *Indus. Engg Chem.,* **60**, (8), 68.
16. Zisman, W. A. (1964). Relation of equilibrium contact angle to liquid and solid constitution. *Contact angle, wettability and adhesion,* Adv. Chem., **43**, American Chemical Society, Washington, 1.
17. Parfitt, G. D. (1967). *J. Oil Colour Chem. Assoc.,* **50**, 822.
18. Crowl, V. T. *Ibid.,* 1023.
    Crowl, V. T. and Malati, M. A. (1966). *Disc. Faraday Soc.,* **42**, 301.
    Titangesellschaft, M. B. H. British Patent 1,157,060 (July 2, 1969).
19. Rechmann, H. (1966). *Farbe Lack,* **72**, 1063.

20. Ashmead, B. V., Bowrey, M., Burrill, P. M., Kendrick, T. C. and Owen, M. J. (1971). *J. Oil Colour Chem. Assoc.*, **54**, 403.
21. Dunn, E. J., Swartz, H. E., Baier, C. H. and Zuccarello, R. K. (1968). *J. Paint Technol.*, **40**, 112.
22. Barksdale, J. (1966). *Titanium—its occurrence, chemistry and technology*, 2nd edition, Ronald Press, New York, 533.
23. Gerteis, R. L. and Elm, A. C. (1971). *J. Paint Technol.*, **43**, 99.
24. Moody, J. R. and Lederer, G. U.S. patent 3,515,566 (2 June, 1970).
25. Santos, P. C. U.S. patent 3,505,091 (7 April, 1970).
26. Laporte Titanium Ltd., Fr. pat., 1,562,951 (11 May, 1969).
27. Foss, W. M. Ger. offen. 1,943,597 (5 March, 1970).
28. British Titan Products Ltd., Fr. pat, 1,579,699 (29 August, 1969).
29. Lederer, G. and Goldsbrough, K., Ger. offen. 2,046,739 (6 May, 1971).
30. Bell, S. H. (1961). Pigments and fillers for use in paints. *Powders in industry*, Society for Chemical Industry, London, 169.
31. Zorll, U., (1969). *Farbe Lacke*, **75**, 1045.
32. Schue, G. K. (1968). *Amer. Paint J.*, **52**, (55), 14.
33. Pellett, G. (1968). *Amer. Ink Mkr.*, **46**, (5), 67.
34. Hund, F. (1966). *Chem.-Ingr-Tech.*, **38**, 423.
35. Scott, J. R. (1961). Fillers and pigments for use in rubber and plastics. *Powders in industry*, Society for Chemical Industry, London, 195.
36. Kruger, H. and Berndt, W. (1968). *Wbl. Papfabr.*, **96**, (14), 517.
37. *Pigmented coating processes for paper and board* (1964). Tappi monograph **28**, Technical Association of the Pulp and Paper Industry, New York.
38. *Paper coating pigments* (1966). Tappi monograph 30, Technical Association of the Pulp and Paper Industry, New York.
39. Muller, P. (1965). *Tappi*, **48**, (8), 55A.
40. Diamond, A. S. (1967). *Tappi*, **50**, (12), 67A.
41. Wood, C. (1965). Xerographic properties of photoconductor binder layers. *Xerography and related processes*. Edited by J. H. Dessauer, and H. E. Clark, Focal Press, New York, 119.
42. Weisbeck, R. (1968). *Chem.-Ingr-Tech.*, **40**, (3), 100.
43. Tereshko, J. W. U.S. patent 3,562,101 (9 February, 1971).

CHAPTER 7

SINTERING AND HOT-PRESSING OF FINE POWDERS: PROPERTIES OF RESULTING MICROSTRUCTURE

SINTERING

General consideration of particle size in sintering

The driving force for sintering is the reduction of surface free energy. Since this energy is high for a powder of high specific surface, from this viewpoint such powders may be expected to show an increased rate of sintering. The geometric changes by which the decrease in surface area occurs during sintering are not at present completely amenable to quantitative description, but among the factors involved are the size- and shape-distribution of the particles, as well as their packing and surface characteristics. However, certain general changes occur and three stages of sintering are recognised (*see* reference 1):

(a) In this first stage there is an increase in the area of interparticulate contact and a rounding of re-entrant angles at the points of contact. Where densification occurs the centres of particles move closer together.

(b) The intermediate stage is considered to have been reached when growing necks between particles meet. The mass may then be thought of as polycrystalline with connected intergranular porosity. As the pores shrink, grain growth occurs.

(c) In the final stages of sintering the pores are discontinuous. These stages blend into each other and each is a gradual process.

The initial stages of sintering lend themselves to theoretical treatment provided a close-packed mass of spherical particles can be assumed. Proposed mechanisms are listed in Table 17.[2] For the intermediate stage the most-used model views grains as regular fourteen-sided figures[3] or closely-related shapes[4] arranged in

121

cubic close packing. A linear relationship is predicted between density and (log time), but grain size is taken as a constant so the model is an approximation only. An improved model has recently been devised which allows for grain growth and assumes that factors pertaining to mass transport, etc., in the first stage persist into the intermediate stage.[5] This leads to a fairly complicated equation, experimental confirmation of which can be achieved by a fairly extensive series of measurements on grain size, grain boundary-pore intersections, density and size of pores and relative numbers of

TABLE 17

MECHANISMS FOR THE INITIAL STAGES OF SINTERING[2]

Mechanism	Effect	m^*	n^*
Surface diffusion	Neck growth only	7	3
Evaporation–condensation		3	1
Viscous flow or plastic flow	Neck growth	2	1
Volume diffusion	and densification	5	2

$* \text{ In } \dfrac{x^m}{a^n} = F(T)t$

where: x is the neck radius between two spherical particles of radius a; T is temperature and t is time.

concave and convex pore surfaces. The proposed model, which has not yet been widely adopted, has the merit that it is capable of providing information on volume and grain boundary diffusion contributions to densification.

A different approach has been to include in the conventional shrinkage equation an empirical relationship found to describe grain growth in a 50 per cent dense compact. The modified equation was discovered to be applicable to some earlier sintering data.[6] It remains to be seen whether these improved models are adequate to represent the sintering behaviour of real powders in which particles may be of widely differing size, etc.

The final stages of sintering, when pores are closed and are removed only by diffusion, are dependent on grain size. Diffusion rates are enhanced at grain boundaries, so pores adjacent to such boundaries are more readily removed than those further away.[1] The nature of the gas filling the pores is also relevant; solubility in the matrix will increase diffusion rates.[7]

The main points in the sintering process where advantage might be obtained by the use of finer powders may be summarised as follows:[8]

(a) in the initial stage, especially where the mechanism is diffusional. Under these conditions the initial shrinkage is proportional to $(1/a^3)^m$ where a is the particle size and m is a constant in the range 0·5–0·4.

(b) if the finer grain size present initially can be maintained to the final stages, pore removal by diffusion will be assisted.

On the other hand, this high initial rate of sintering may lead to uneven grain growth or uneven densification with the formation of large pores which are very difficult to remove.[1]

A more pragmatic approach to the study of sintering has been by transmission electron microscopy. Using principally magnesia, this has led to a morphological description of the sintering and grain growth processes, distinguishing between events at 900°C, when surface diffusion is thought to be the dominant transport mechanism, and 1 100°C, when vapour phase transport seems more likely.[9]

Sintering of submicrometre powders—experimental

Experimentally it has been known for some time that finer particles sinter more rapidly than those which are larger.[10–12] Quantitative studies on particles in the size range of tens to thousands of micrometres have been used to help in establishing which mechanism is responsible for sintering.[13] The qualitative relation between size and sintering rate has been extended to submicrometre particles sintering in a compact,[14] but since submicrometre powders do not contain spheres of sufficiently uniform size, a quantitative relation has not emerged. When attempts have been made to examine neck growth rates between pairs of submicrometre spherical particles heated in an electron microscope, the rates were not very reproducible and not enough data were gathered to show accurately and conclusively the relation between size and sintering rate for these particles.[15]

It must be stressed that the models used cover the sintering of something approaching a close packed lattice of equal sized particles and cover only the first few per cent of shrinkage. Some attempt has been made to apply the model to the later stages of sintering of what must initially have been a relatively low density compact.[16] With very fine powders of hard materials it may be difficult to reach a suitable degree of cold densification prior to sintering.

Demonstration of the effect of particle size on the later stages of sintering is much more difficult. In the intermediate stage the particle size of a series of samples of the same material can be considered to be the same at the onset of that stage. It can then be shown that the grain size at a fixed degree of densification is independent of the temperature of sintering.[4] In the final stage of sintering, considerable variations in behaviour are found. These have been attributed to discontinuous pore growth,[17] the effect of atmosphere, etc., but may very well reflect, in part, the preparation of the starting material. A powder prepared by thermal decomposition of a carbonate or hydroxide may hold tenaciously in some micropores traces of carbon dioxide or water which will be expelled only under the extreme conditions required for near-complete densification. For example, mass spectrometry and other techniques show water and carbon dioxide to be present in hot-pressed alumina.[18]

HOT-PRESSING

Densification is facilitated when pressure is applied during the sintering operation; a lower temperature may be used and there is less grain growth.[19,20] The driving forces for densification are surface energy and applied pressure and models for the three stages of densification have been devised.[21] When very high pressures are used (a few GN/m^2) the temperature required to reach essentially complete densification may be reduced to the point where virtually no overall grain growth occurs during densification. This state has been achieved with submicrometre barium titanate, magnesia, alumina, chromia and tungsten and fine grained silicon carbide and tungsten boride.[22] In order to take the fullest advantage of the fine microstructure which can be produced, fine powders are required as starting materials.

Data on the kinetics of the initial stages of hot-pressing are scanty, but it seems that the primary mechanisms of densification are boundary sliding and particle fracture with a variable amount of plastic flow.[23] In the later stages of densification, if the pressure is high and the yield stress low (e.g. for lead[24] or nickel oxide[25]) almost all the densification might be by plastic flow, but at low pressures with high yield stress (*e.g.* for alumina[26]), diffusional mechanisms may be expected to predominate.

Observed rates of densification on hot-pressing increase with decreasing particle size for hard materials such as alumina[26] and tungsten carbide[27] where diffusion control is to be expected. This leads to a kinetic relationship of the same form as that for plastic flow of a Bingham solid,[26] and for some time it was incorrectly thought that plastic flow occurred on hot-pressing brittle materials.[28] In the hot-pressing of silica, where plastic or viscous flow can be expected to occur, the influence of particle size is quite insignificant.[29]

Thus, as in pressing and sintering, the hot-pressing of a hard material in a finer particle size may lead to the production of a body with a finer microstructure.

SINTERING OF POLYCOMPONENT MIXTURES

Just as a finer single component powder will densify at a lower temperature than one in which the particles are larger, so a mixture of two fine powders may be expected to react at a lower temperature than the same powders in a coarse form. However, as compared with the sintering of a single component, solid–solid reactions may be complicated. The diffusion rates of the components may differ quite widely so that swelling or shrinkage of particles may occur, as well as severe deformation of the interparticle neck, and this may change the mechanism in subsequent stages.[30] For example, in the sintering of ferric oxide–magnesium oxide and ferric oxide–nickel oxide mixtures, the initial neck growth is normal, but later the neck becomes distorted because of interdiffusion.[31]

In a properly blended mixture a reduction in particle size increases the number of contacts between the phases, and reduces the distance over which diffusion must take place before the composition becomes uniform. These effects, together with the accelerated sintering due to smaller particles, make fine powders desirable for solid–solid reactions. However, each case must be treated on its merits. The lower temperature densification which may be achieved by hot pressure is especially valuable when one of the components is slightly volatile.

GRAIN GROWTH INHIBITION AND
SINTERING ADDITIVES

It is well established that in many materials impurities segregate to grain boundaries. The concentration of such impurities is usually insufficient to exert detectable influence on sintering behaviour. When the finest powders are compacted, commonly-observed anion impurities such as Cl, S, F, OH and CO_2 may result from adsorption onto the surface of the powder before compaction.[32] When a larger amount of an impurity (or additive) which segregates to the grain boundary is present, the boundary may be rendered much less mobile or effectively anchored. In this way, grain growth may be prevented or reduced with benefit to the microstructure of the resulting compact. An addition of calcia to thoria has been shown to impede grain boundary movement while at the same time maintaining a high flux of vacancies from pores to grain boundaries.[33] With such grain growth inhibitors, care is needed to ensure that a continuous film of additive, which might reduce mechanical strength, is not formed at the grain boundary.[34]

Other additives form small amounts of liquid phase at grain boundaries. This may lubricate particle rearrangement, and later in densification may form a transport medium. Such a material is lithium fluoride added to magnesia[35] which considerably increases the rate of sintering. Other sintering aids, such as titania in alumina[36] are believed to concentrate on grain surfaces and to enhance diffusion. With these additives care is needed to ensure that increased grain growth does not occur.[34]

The possible influence of additives used for these or any other purposes on the properties of the microstructure produced must always be borne in mind.

MICROSTRUCTURE AND
PROPERTIES

It has been shown that the use of finer powders for sintering or hot-pressing should result in finer grain size or, perhaps, higher relative densities. We shall now examine the effects of these microstructural parameters on properties.

Optical properties

Virtually complete densification is required to make ceramics transparent or translucent.[37] This is because, at a phase boundary, scattering is approximately proportional to the square of the difference in the refractive indices of the two phases. Grain boundaries, on the other hand, produce only limited scattering, the extent of which varies with the degree of anisotropy of refractive index existing in the crystal, being thus zero for a material with a cubic unit cell such as magnesia, 0·008 for alumina and 0·014 for beryllia.[38] Thus, fully dense magnesia should be transparent whilst alumina should be more translucent than beryllia. When the crystallite size becomes markedly less than the wavelength of light, scattering at grain boundaries will become less and the material will appear more translucent, provided that it is fully dense.

Transparent alumina, which is a commercial material, contains magnesia added as a grain growth inhibitor.[39]

Mechanical properties

When the grains are not very small the strength of a polycrystalline ceramic body can be described as a function of grain size and porosity by such empirical relations as:[40]

$$S = kG^{-a}e^{-bp}$$

where: k, a and b are empirical constants,
         G is the grain size (>1 μm) and p the porosity.

This relation is similar to the Petch relationship for metals. At very fine grain sizes, rupture strengths are often not as high as predicted by this type of relationship, and strength may even decrease.[33,41,42] This can be explained in terms of contamination of the grain boundaries, but it is possible that surface damage is now the strength-controlling factor as it is in amorphous solids where the grain size has become vanishingly small. Under certain conditions, usually when strengths are high, the state of the surface is already known to be important, but the effect is much greater in single crystal sapphire[43] than in polycrystalline alumina.[44] However, it must be said that vacuum hot-pressing does sometimes produce very strong ceramics[45] where the empirical relations, like that given above, may be applicable and grain sizes are still very small. These occasions seem to be rare and it is fairly generally recognised that there is likely to be an effective maximum beyond which further reduction in grain size will not increase strength.

To produce a uniform and strictly flat surface on a pressed and sintered or hot-pressed body, a fine grain size and freedom from discontinuous grain growth during densification is needed. Such a surface is required as substrate if it is desired to deposit thin films of good quality and uniformity for use as electronic devices. Because of its high thermal conductivity and low electrical conductivity, alumina is preferred to glass as a substrate material whenever appreciable power is dissipated in the circuit. For thin film substrates the surface must be flat to within about 120 nm. Grains must be smaller. To achieve this, the finest particle size alumina must be hot-pressed and care taken to avoid undesirably high surface concentrations of alkali, iron etc., resulting from segregation.[46] Most methods for the preparation of fine alumina lead to one or more of the metastable forms (γ, δ, etc.) which, on hot-pressing, convert to the α-form with considerable grain growth. However, submicrometre particle size α-alumina may be produced by careful calcination of gibbsite or boehmite. Reactive hot-pressing of these starting materials leads to high density α-alumina with grain sizes not much greater than a micrometre.[47] It may, therefore, be very difficult to produce alumina substrates of the required degree of smoothness unless some kind of grinding process is capable of producing acceptable surfaces. For thick-film technology the specification for the substrate surface is considerably less demanding.

Magnetic properties

The 'structure sensitive' magnetic properties are the only ones of interest here. Such principal properties are:[48]

—permeability
—magnetic loss due to hysteresis, eddy currents, etc.
—coercivity

The main uses of ceramic magnetic materials lie in permanent magnets and in magnetic cores operating at a wide range of alternating current frequencies.

In a permanent magnet it is desired to retain as high a degree of magnetisation as possible after the external field is removed. Reversal is made more difficult by using a material which shows structural magnetic anisotropy. A suitable structure is that of barium (or strontium) ferrite, $BaFe_{12}O_{19}$, which is often used for permanent magnets,[48] but coercivities achieved in practice are much less than the anisotropy fields arising from the crystallographic structure of the ferrite used. This is due to the formation of domains within the grains so that part of the demagnetisation is due to wall movements.[49] The critical grain size above which domain walls appear is about

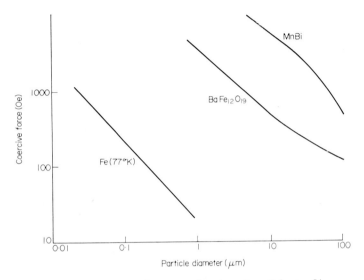

Fig. 25.   Coercive force as a function of particle size.[51]

1·3 μm. To form a microstructure of less than this size, hot-pressing of a submicrometre size powder is carried out.[50] Figure 25 shows the relationship between grain size and coercive force for a selection of magnetic materials.[51] The saturation magnetisation of barium ferrite[52] has been shown to decrease somewhat at a grain size less than about 0·5 μm. This has been explained as due to the influence of a surface layer of properties different from the bulk.

The main requirement for magnetic cores operating under alternating current conditions is for high permeability with low magnetic losses. Losses due to hysteresis may be reduced by reducing the coercivity (field strength at zero induction). The same microstructural factor which increases permeability reduces coercivity. This factor is freedom from pores, inclusions and grain boundaries which hinder domain boundary movement.[50] A homogeneous microstructure is required with grain sizes larger than domain size (*i.e.* several micrometres in diameter). At frequencies up to about 3 MHz, manganese zinc ferrite is usually used for inductors. Above this figure a nickel zinc ferrite is used because of its higher resistivity and consequent lower eddy current losses. In high flux density applications, such as television line output transformers, the change is made at lower frequencies (~ 0.5 MHz).[49]

At higher (microwave) frequencies, for example in radar, there are several uses for ferrites. For devices capable of handling high levels of peak power a grain size below about 1 μm is desirable. The grain boundaries thus introduced appear to increase the critical value of the field beyond which spin wave excitation and instability set in.[53] A range of ferrites has been prepared by spraying mixed nitrate solutions into a flame, and hot-pressing the oxides thus formed.[54,55] The ferrite particles were about 0.02 μm diameter, but because the sintering temperatures were lower, dielectric losses were high owing to inhomogeneities not smoothed out on sintering. Under these conditions a higher standard of uniformity is demanded of the initial powder.[56] The best high power ferrites are still made by grinding coarser material and hot-pressing.[57,58] As an example of what might be achieved, reduction of grain size from 15 μm to 2 μm increases by a factor of about ten the peak power which may be handled.[50]

In computer memories ferrites with rectangular hysteresis loops are used and switching is accomplished by the changeover from one remanent state to the other. The relationship of microstructure to the switching mechanism is not fully understood, but it is known that a lithium ferrite of about 0.7 μm grain size switches at about twice the speed of larger grain size material, with little or no increase in coercive force.[59] It has also been suggested that ultra-fine nickel ferrite can be used to produce memory cores with a signal output more than twice that of conventional cores.[60]

In magnetic materials with a grain size below about 10–15 nm, thermal energy can be sufficient to disturb the parallel arrangement of atomic moments associated with a single domain.[44] This can result in a reduction to zero of the magnetisation of an assembly of such particles.[63] A suspension of subdomain magnetic particles in a liquid behaves similarly.[61] However, when single particles form chains, individual particles, although below the critical size, do not show this behaviour[62] which is termed superparamagnetism.

Electrical properties

In ferroelectric materials for use as piezoelectrics or in capacitors, a high relative density is required. Particle size effects are seen only when the grain size is not much greater than the domain size (*i.e.* $\sim 1~\mu m$). In this size range hysteresis loops become narrower and remanence decreases as the particle size decreases. Many fine-grained barium titanates contain donor ion additives and grain growth inhibitors.[64] These are especially valuable for use in capacitors. The use of pure, submicrometre grained barium titanate is severely restricted at present[21,65]. Ferroelectric materials such as lead titanate–lead zirconate proposed for use in electro-optics show significantly different properties when the grain size falls below about 1–2 μm when the optical transmittance becomes much higher. It has been suggested that this is due to a relative absence of domains.[51,66]

Possible future developments

Where the practical aim of sintering or hot-pressing is to achieve high or complete densification with control of grain growth the particle size of the starting material must be less than the required grain size. Provided a suitable grain growth inhibitor can be found. sintering or hot-pressing may then produce the required micro-structure. Hot-pressing is somewhat restricted at present by a lack of readily-available non-reactive materials for the construction of dies, but the technique is well suited for the production of high density compacts. In sintering, on the other hand, once a large pore has been produced it may be difficult, or virtually impossible, to remove it. Uniform densification is required. Most of the aggregates present in a fine powder are easily broken down on pressing, but considerable work is needed to study the microstructural changes which occur during the hot-pressing or sintering of practical systems, the sizes and distribution of pores, etc.

It will be surprising if future work on electrical, magnetic and optical properties of compacts does not call for finer grain structures than are now available, as well as for improvements in the compositional homogeneity. It will then be necessary to prepare more perfect fine powders and greater use will probably be made of preparative routes involving gas phase reactions.

REFERENCES

1. Coble, R. L. and Burke, J. E. (1963). Sintering in ceramics. *Prog. ceram. Sci.,* **3.** Edited by J. E. Burke, Pergamon, Oxford, 197.
2. Kuczynski, G. C. (1949). *Trans. Amer. Inst. Mining metall. Engrs,* **185,** 169.
3. Coble, R. L. (1961). *J. appl. Phys.,* **32,** 787, 793.
4. Coble, R. L. and Gupta, T. K. (1967). Intermediate stage sintering. *Sintering and related phenomena.* Edited by G. C. Kuczynski, N. A. Hooton and C. F. Gibbon, Gordon and Breach, New York, 423.
5. Johnson, D. L. (1970). *J. Amer. ceram. Soc.,* **53,** 574.
6. Moriyoshi, Y. and Komatsu, W. (1970). *J. Amer. ceram. Soc.,* **53,** 671.
7. Coble, R. L. (1962). *J. Amer. ceram. Soc.,* **45,** 123.
8. Kuhn, W. E. (1963). Consolidation of ultrafine particles. *Ultrafine particles.* Edited by W. E. Kuhn, Wiley, New York, 41.
9. Stringer, R. K., Warble, C. E. and Williams, L. S. (1969). Phenomenological observations during solid reactions. *Kinetics of reactions in ionic systems.* Edited by T. J. Gray and V. D. Frechette, Plenum Press, New York, 53.
10. Kuczynski, G. C. (1950). *J. appl. Phys.,* **21,** 632.
11. Moser, J. B. and Whitmore, D. H. (1960). *J. appl. Phys.,* **31,** 488.
12. Wilder, D. R. and Fitzsimmons, E. S. (1955). *J. Amer. ceram. Soc.,* **38,** 66.
13. Coble, R. L. (1958). *J. Amer. ceram. Soc.,* **41,** 55.
14. Johnson, D. L. and Cutler, I. B. (1963). *J. Amer. ceram. Soc.,* **46,** 541, 545.
15. Vahldiek, F. W., Swihart, D. E. and Mersol, S. A. (1967). Sintering study of Al_2O_3 and ZrO_2 submicron particles by electron microscopy. *Sintering and related phenomena.* Edited by G. C. Kuczynski, N. A. Hooton and C. F. Gibbon, Gordon and Breach, New York, 617.
16. Kothari, N. C. (1965). *J. nucl. Mater.,* **17,** 43.
17. Gupta, T. K. and Coble, R. L. (1968). *J. Amer. ceram. Soc.,* **51,** 521, 525.
18. Rice, R. W. (1968). Strength and fracture of dense MgO. *Ceramic microstructures.* Edited by R. M. Fulrath and J. A. Pask, Wiley, New York, 579.
19. Vasilos, T. and Spriggs, R. M. (1965). *Proc. Brit. ceram. Soc.,* **3,** 195.
20. Coble, R. L. (1967). Mechanisms of densification during hot pressing. *Sintering and related phenomena.* Edited by G. C. Kuczynski, N. A. Hooton and C. F. Gibbon, Gordon and Breach, New York, 329.
21. Coble, R. L. (1970). *J. appl. Phys.,* **41,** 4798.
22. Kalish, D. and Clougherty, E. V. (1969). *Bull. Amer. ceram. Soc.,* **48,** 570.
23. Felten, E. J. (1961). *J. Amer. ceram. Soc.,* **44,** 381.
24. Westerman, F. E. and Carlson, R. G. (1961). *Trans. Amer. Inst. Mining metall. Engrs,* **221,** 649.
25. Spriggs, R., Brissette, L. A. and Vasilos, T. (1964). *Bull. Amer. ceram. Soc.,* **43,** 572.
26. Rossi, R. C. and Fulrath, R. M. (1965). *J. Amer. ceram. Soc.,* **48,** 558.
27. Jackson, J. S. and Palmer, P. F. (1960). Hot pressing refractory hard materials. *Special ceramics.* Edited by P. Popper, Heywood, London, 305; Jackson, J. S. (1961). *Powder Metall.,* (8), 73.
28. Murray, P., Rodgers, E. P. and Williams, A. E. (1954). *Trans. Brit. ceram. Soc.,* **53,** 474.
29. Vasilos, T. (1960). *J. Amer. ceram. Soc.,* **43,** 517.
30. Kuczynski, G. C. (1967). Sintering in multicomponent systems. *Sintering and related phenomena.* Edited by G. C. Kuczynski, N. A. Hooton, and C. F. Gibbon, Gordon and Breach, New York, 685.
31. Venkatu, D. A. and Kuczynski, G. C. (1969). Sintering of two-component

oxide systems with compound formation. *Kinetics of reactions in ionic systems.* Edited by T. J. Gray and V. D. Frechette, Plenum Press, New York, 316.

32. Leipold, M. H. and Blosser, E. R. (1970). The role of composition in ultrafine-grain ceramics. *Ultrafine grain ceramics.* Edited by J. J. Burke, N. L. Reed and V. Weiss, Syracuse University Press, 99.
33. Jorgensen, P. J. and Schmidt, W. G. (1970). *J. Amer. ceram. Soc.,* **53,** 24.
34. Hall, R. C. (1968). *Bull. Amer. ceram. Soc.,* **47,** 251.
35. Hart, P. E., Aitkin, R. B. and Pask, J. A. (1970). *J. Amer. ceram. Soc.,* **53,** 83.
36. Bagley, R. D., Cutler, I. B. and Johnson, D. L. (1970). *J. Amer. ceram. Soc.,* **53,** 136.
37. Gardner, W. J., McClelland, J. D. and Richardson, J. H. (1965). Hot pressed translucent oxides. *Modern ceramics—some principles and concepts.* Edited by J. E. Hove and W. C. Riley, Wiley, New York, 215
38. Winchell, A. N. and Winchell, H. (1964). *The microscopical characters of artificial inorganic substances: optical properties of artificial minerals,* Academic Press, New York, 58, 60.
39. Kim, Y. S. and Monforte, F. R. (1971). *Bull. Amer. ceram. Soc.,* **50,** 532.
40. Knudsen, F. P. (1959). *J. Amer. ceram. Soc.,* **42,** 376.
41. Spriggs, R. M., Brissette, L. A. and Vasilos, T. (1964). *Bull. Amer. ceram. Soc.,* **43,** 572.
42. Spriggs, R. M., Brissette, L. A. and Vasilos, T. (1963). *J. Amer. ceram. Soc.,* **46,** 508.
43. Mallinder, F. P. and Proctor, B. (1966). *Proc. Brit. ceram. Soc.,* **6,** 9.
44. Steele, B. R., Rigby, F. and Hesketh, M. C. (1966). *Ibid.,* 83.
45. Vasilos, T. and Spriggs, R. M. (1967). Microstructure in oxides. *Sintering and related phenomena.* Edited by G. C. Kuczynski, N. A. Hooton and C. F. Gibbon, Gordon and Breach, New York, 301.
46. Buchanan, R. C. (1968). *Ceramic Age,* **84,** 60.
47. Matkin, D. I., Munro, W. and Valentine, T. M. (1971). *J. Mater. Sci.,* **6,** 974.
48. Brockman, F. G. (1968). *Bull. Amer. ceram. Soc.,* **47,** 186.
49. Snelling, E. C. (1970). *Proc. Brit. ceram. Soc.,* **18,** 87.
50. Stuijts, A. L. (1968). Microstructural considerations in ferromagnetic ceramics. *Ceramic microstructures.* Edited by R. M. Fulrath and J. A. Pask, Wiley, New York, 443.
51. Mountvala, A. J. (1970). Electrical and magnetic behaviour of ultrafine grain ceramics. *Ultrafine grain ceramics.* Edited by J. J. Burke, N. L. Reed and V. Weiss, Syracuse University Press, 367.
52. Torkar, K. and Fredriksen, O. (1959). *Powder-Met.,* **4,** 105.
53. Borghese, C. and Roveda, R. (1969). *J. appl. Phys.,* **40,** 4791.
54. Malinofsky, W. W. and Babbit, R. W. (1961). *J. appl. Phys.,* **32,** 237S: (1964), **35,** 1012.
55. Malinofsky, W. W., Babbit, R. W. and Sands, G. C. (1962). *J. appl. Phys.,* **33,** 1206.
56. De Lau, J. G. M. (1970). *Bull. Amer. ceram. Soc.,* **49,** 572.
57. Blankenship, A. C. and Hunt, R. L. (1966). *J. appl. Phys.,* **37,** 1066.
58. Wantuch, E. and Lepore, D. A. (1966). *Ibid.,* 1079.
59. Babbit, R. W., Sands, G. and Dunlop, A. (1969). *J. appl. Phys.,* **40,** 1455.
60. Goetzel, C. G. and Steinberg, M. A. (1966). A new technology based on submicron powders. *Modern developments in powder metallurgy,* **2,** edited by H. H. Hausner, Plenum Press, New York, 194.
61. Kaiser, R. and Miskolczy, G. (1970). *J. appl. Phys.,* **41,** 1064.
62. Chock, E. P. (1970). *Jap. J. appl. Phys.,* **9,** 410.

63. Eibschutz, M. and Shtrickman, S. (1968). *J. appl. Phys.,* **39,** 997.
64. Jaffe, B. (1968). Effect of microstructure on ferroelectric and piezoelectric properties. *Ceramic microstructure.* Edited by R. M. Fulrath and J. A. Pask, Wiley, New York, 475.
65. Egerton, L. and Koonce, S. E. (1955). *J. Amer. ceram. Soc.,* **38,** 412.
66. Haertling, G. H. (1970). *Bull. Amer. ceram. Soc.,* **49,** 564.

CHAPTER 8

OTHER PROPERTIES OF FINE POWDERS

AGGLOMERATION IN FINE POWDERS

As prepared, the particles in a fine powder are in a more or less agglomerated state with varying quantities of (usually) air and water between the particles. Forces causing primary particles to stick together have been classified as:[1]

—solid bridging;
—liquid bridging;
—attraction between particles;
—interlocking.

Solid bridges can form as a result of sintering, solid diffusion or chemical reaction, all of which are more likely to occur at elevated temperatures. Ex-solution of soluble matter can form solid bridges at room temperature.

Liquid bridging results from the presence between the individual particles of bulk liquid. At normal humidities fine powders may contain considerable amounts of water. For example, a fine alumina (specific surface 250 m^2) has been shown to contain about 11 per cent by weight of water. The forces existing inside an agglomerate in the presence of liquid have been considered[2] and four stages distinguished, depending on the amount of liquid. The first is that in which insufficient liquid is present to completely fill spaces in the agglomerate and liquid is concentrated in saddles between particles. The second is when gas is still present, but the liquid forms a continuous network. In the third, the space inside the agglomerate is just filled with liquid. In these three stages there is negative capillary pressure in the filled saddles, and, where spaces are only partly filled, surface tension at the liquid–gas interface also contributes to the bonding force between grains. In the fourth

135

stage, liquid coats the agglomerate and surrounds it in a droplet and surface tension forces on the outside of the drop alone keep the particles together. Once a liquid bridge is established any evaporation of liquid reduces the radii of curvature of the liquid–gas interface and thus increases the forces holding the particles together so that they approach each other more closely.

In addition, it is possible for the liquid–particle interface to show appreciable strength. This may occur because of surface irregularities or it may result from some form of bonding between the adsorbed liquid layer and the surface of the particles. In some cases, at least, it is likely that hydrogen bonding can occur between a polar liquid, such as water, and surface groupings, such as hydroxyl. The part which may be played by such bonding in bridging operations between particles seems to have received insufficient recognition, but can be seen in the thixotropy exhibited by dilute suspensions of some fine powders (see page 136).

Forces of interparticle attraction are of two types, electrostatic and Van der Waal's. Electrostatic forces arise through charging by contact with charged particles, or by friction. Many fine powders prepared by high temperature processes are exposed to flames or plasma where charged particles abound. Often, electrostatic precipitators[3] are used for the collection of fine powders. While a charged particle soon loses a portion of its charge by interaction with cosmic radiation, etc., giving a Boltzman distribution of charge dependent on particle size,[4] once electrostatic agglomeration has occurred, particles do not separate without work being carried out. These electrostatic forces affect the shape of the agglomerate. Highly-charged particles form chains while uncharged particles tend to form spherical clusters.[5]

Van der Waal's forces arise from electron motion within the atoms and protrude beyond the surface of the particle. These forces, given approximately by $(W/L^2)[(r_1 r_2)/(r_1 + r_2)]$ where r_1 and r_2 are the radii of two spherical particles the centres of which are L apart, and W is a constant, are appreciable only where r_1, r_2 and L are small.

The formation of agglomerates by particle interlocking can occur with fibrous or plate-shaped particles, but not with spheres.

For fine particles (diameter well below 1 μm) the most important agglomerating forces are likely to be Van der Waal's, together with those due to liquid bridging. The tap density of such powders may be as low as 1 per cent of the theoretical density, so particle bridging and extensive void formation must occur. While there is, as yet, no theoretical treatment of particle–particle adhesion, the adhesion of particles to a plane surface has been treated more or less quantitatively.[6]

To give an idea of the relative magnitude of the forces concerned, for 100 nm particles separated by about 3 nm, Van der Waal's forces will be equivalent to about 1 atmosphere, while capillary forces due to liquid bridging may be of the order of a hundred times this value.[1]

Methods of agglomeration

Although fine powders are, to some extent, agglomerated in the course of preparation, bulk densities are low and handling not easy. Further agglomeration can be produced by using the densifying forces which come into play when fluid added to a fine powder is subsequently evaporated (*see* previous section). Larger agglomerates are less stable, once formed, than those formed spontaneously in fine powders, and when the material is inherently hard it is often necessary either to allow a residue of fluid to remain or to add a low viscosity binder. These principles are employed in such agglomerating processes as granulation and spray drying. The gelling of spheres of sol also leads to the formation of easily-handled agglomerates.

When applied to very fine powders, such as those mentioned above, in which the tap density is about 1 per cent of theoretical, agglomeration by mixing with a fluid and granulating may easily increase the density to, say, 20 per cent of theoretical.

FLOW PROPERTIES OF FINE POWDERS

Fine powders do not flow easily. Agglomerates move rather than individual particles and there is a continuous process of de-agglomeration and re-agglomeration. The flow of large particles can be successfully treated by the principles of soil mechanics, but the factors involved in a quantitative description of such flow cannot at present be defined adequately for fine material.

Small amounts of fine powders can often be used as additives to improve the flow properties of a range of powdered insecticides, chemicals, pharmaceuticals, etc. It appears that during mixing the agglomerates of fine powder are broken down, and the fine particles coat the coarser particles of the main constituent and prevent these sticking together. Especially valuable in this connection are likely to be coated hydrophobic forms of fine powders.[7] A related use for fine powders is to completely absorb small quantities of liquids which are thus effectively converted to free-flowing powders which may be readily incorporated into a large mass of dry powder. This may be of value for flavouring, syrups, etc. Liquids such as alcohols, mineral spirits, sulphuric acid and water can be added as free-flowing powders when absorbed on fine powders.[8] Similarly, damp or sticky solids, otherwise difficult to handle, may be converted to a free-flowing powder.[9]

PACKING OF FINE POWDERS

The packing of fine powders under their own weight is expressed as the tap density. This density may be expected to depend on the shape, absolute size and size distribution, and surface characteristics of the ultimate particles, as well as the state of agglomeration of the powder. Bridging and arching of agglomerates may lead to very low tap densities. Thus, alumina prepared by a condensation process may have a tap density of 0·03 g/cc. Other samples similarly prepared, and with a similar surface area, may, on the other hand, show a tap density fourteen times the above value, due to differences in the degree of agglomeration. When one particular preparation of alumina or ferric oxide is progressively reduced in size, there is a progressive decrease in tap density.[10]

PRESSING

When a material is pressed, whether unidirectionally or in two directions in a die, or isostatically when pressure is transmitted by a fluid, compaction can occur by two means:

  (a) by particle rearrangement by sliding;
  (b) by deformation and/or fracture.

In the early stages of pressing, the forces resisting compaction are those holding agglomerates together augmented by friction between particles and with the die wall. As the pressure is increased, the force acting at interparticle contacts may become sufficient to produce deformation or fracture. Hard brittle materials will tend to fracture at high pressures,[11] soft ductile materials to deform.

On compacting a fine powder, a considerable reduction in volume occurs at very low applied pressure (a few hundred kN/m^2).[10] At these pressures only the relatively weak forces of bridging and agglomeration considered above will be overcome. At higher applied pressures the density of a pressed pellet of alumina, for example, increases with increased particle size of the starting powder,[12,13] and is little affected by the use of die lubricants, although such lubricants improve the strength of the pressed pellet.[11] This strength will, to a considerable extent, be a reflection of the bonding of the material undergoing pressing. Thus, crystals with layer structures (such as graphite or boron nitride) dependent on Van der Waal's forces between the layers, produce too few strong bonds to give a readily-handled pellet. The same is true of hard substances such as silicon carbide in which there is little particle deformation and, consequently, the interparticle contact area remains small. When pressing a fine powder, the surface roughness of die and plunger may be comparable with the particle size, but it is not clear what the effect of this might be.

Whilst it might be expected that this type of sliding and rearrangement would be influenced by surface films present on the fine powder, such influence appears to be slight; isostatic compression of alumina powder carrying various adsorbates showed no significant differences in compaction behaviour.[14] With coarser powders surface contamination exerts a complex effect on compaction and adhesion.[15]

THERMAL CONDUCTIVITY

The thermal conductivity of an agglomerated powder in a gas can be calculated satisfactorily by assuming that heat is transferred by conduction through the solid, by conduction through the gas and by particle to particle radiation.[16] For a fine powder, conduction is small owing to the small contact areas between particles, although these areas may be somewhat increased by the presence of adsorbed layers or preferential condensation of liquid into the saddles between particles. Conductivity is independent of gas pressure as long as the mean free path of the gas molecule is less than the interparticle spacing.

Powder-filled but evacuated spaces have been used to insulate large liquid-gas containers and other low-temperature devices where hard vacuum is difficult to maintain.[16]

When the particles are not agglomerated, as in aerosols, the individual particles are in motion and move down a temperature gradient.[17]

DIELECTRIC LOSS AND SURFACE PROPERTIES

The dissipation factor of dried submicrometre alumina is unaffected below 10^5 Hz by adsorption of a variety of gases such as helium, oxygen, nitrogen and carbon dioxide. Adsorption of water vapour greatly increased the loss due, it is thought, to leaching out of adsorbed electrolyte. Since replacement of adsorbed water is thought to be one of the functions of a binder in pressing, the change in dielectric loss can be used to assess the efficiency of binders.[18]

GAS–SOLID CHEMICAL REACTIONS— COMBUSTION, ETC.

The very large surface area associated with a fine powder means that the actual reaction between the surface and a gas can proceed at an increased rate. It does not, of course, follow that this actual reaction is the rate-determining stage. Thus, when the reactant gas is very dilute, gaseous diffusion to the solid reactant may be the controlling step. Or, again, when the product is solid and deposits as a coherent coating on the surface of the solid reactant, the controlling step may be diffusion through the coating. Such a state of affairs is found in the room temperature oxidation of a number of metals[19] where there is a limiting thickness of the product layer beyond which reaction is effectively stopped. However, since such a layer is typically about 2 nm thick, a substantial proportion of a fine metal powder in which particles are, say, a few tens of nm diameter, will in fact be oxidised before the oxide layer is thick enough to be protective. If the oxidation is exothermic, as is usually the case, then a temperature rise may occur and the degree of protection will be reduced or may cease altogether.

An applied surface coating of polymer has been used to stabilise small pyrophoric iron particles.[20] Some other pyrophoric materials, such as nitrides, carbides and borides, can be stabilised by the formation of a thin layer of oxide providing that the resulting degree of contamination is acceptable.

Much work has been carried out in defining the conditions under which various materials can be expected to explode when the reaction is exothermic.[21,22,23] There is a lower concentration limit for explosion below which insufficient heat is generated to cause the reaction to run away and an upper limit beyond which the concentration of reacting gas is insufficient to sustain reaction.[24] With larger particles, a decrease in size reduces the minimum energy required to produce ignition and increases the rate of reaction and associated pressure increase due to heating of any inert gas (*e.g.* nitrogen in an air explosion). However, there are indications that continued reduction in particle size may not increase the explosion risk when particles are a few micrometres in diameter.[25] This may be due to some other factor becoming rate-controlling.

Dusts of most metals, alloys, suboxides, carbides, hydrides and nitrides which react exothermically with oxygen exhibit pyrophoricity at room temperature. Such dusts must usually be of submicrometre size, but larger particles of uranium, uranium hydride and zirconium are pyrophoric.[23] When the material is capable of reacting exothermically with moisture the risk of explosion is increased.[26]

Powders of non-combustible materials, especially alkali metal salts, although of diameter of some tens of micrometres and thus larger than those under consideration here, can be used to extinguish flames by acting to extract energy and thus reduce the temperature[27] or by limiting the number of available free radicals.[27] Detonations can also be suppressed by such powders, but the mechanisms involved seem to be somewhat different.[28] Difficulties associated with the dispersion of very fine particles can be overcome by the use of materials which decompose in the flame and in so doing decrepitate, thus markedly increasing the available surface area and producing particles of about one micrometre in diameter.[29]

142   FINE POWDERS—PREPARATION, PROPERTIES AND USES

REFERENCES

1. Rumpf, H. (1962). The strength of granules and agglomerates. *Agglomeration.* Edited by W. A. Knepper, Interscience, New York, 379.
2. Newitt, D. M. and Conway-Jones, J. M. (1958). *Trans. Instn. chem. Engrs,* **36,** 422.
3. White, H. J. (1963). *Industrial electrostatic precipitation,* Pergamon, Oxford.
4. Whitby, K. T. and Peterson, C. M. (1965). *Indus. Engg Chem: Fund.,* **4,** 66.
5. Dallavalle, J. M., Orr, C. and Hinkle, B. L. (1954). *Brit. J. appl. Phys.,* **3,** S.198.
6. Krupp, H. (1967). *Adv. Colloid Interf. Sci.,* **1,** 111.
7. Klein, K. (1968). *Chem-Techn Industrie,* **64,** 849.
8. *Cabosil—how to use it, where to use it.* Cabot Corp., Boston, Mass., 1968.
9. Wilcox, J. D. and Klein, J. M. (1971). *Powder Technology,* **5,** 19.
10. Oudemans, G. J. (1965). Compaction of dry ceramic powders. *Science of Ceramics,* **2,** edited by G. H. Stewart, Academic Press, London, 133.
11. Cooper, A. R. and Eaton, L. E. (1962). *J. Amer. ceram. Soc.,* **45,** 98.
12. Kuhn, W. E. (1962). Consolidation of ultrafine particles. *Ultrafine particles.* Edited by W. E. Kuhn, Wiley, New York, 41.
13. Bruch, C. A. (1967). *Ceram. Age,* **83,** (10), 44.
14. Sturgis, D. H. and Nelson, J. A. (1969). *J. Amer. ceram. Soc.,* **52,** 286.
15. Boehme, G., Krupp, H., Rabenhorst, H. and Sandstede, G. (1962). *Trans. Instn. chem. Engrs,* **40,** 252.
16. Orr, C. (1966). *Particulate technology,* Macmillan, New York, 229.
17. Waldmann, L. and Schmitt, K. H. (1966). Thermophoresis and diffusiophoresis of aerosols. *Aerosol Science.* Edited by C. N. Davies, Academic Press, New York, 137.
18. Mountvala, A. J., Onoda, G. Y. and Onesto, E. J. (1971). *Bull. Amer. ceram. Soc.,* **50,** 170, 627.
19. Russell, L. E. (1967). *Powder Metall.,* **10,** (20), 239.
20. Robbins, M., Swisher, J. H., Gladstone, H. W. and Sherwood, R. C. (1970). *J. electrochem. Soc.,* **117,** 137.
21. Jacobson, M., Nagy, J., Cooper, A. R. and Ball, F. J. (1960). *Explosibility of agricultural dusts,* U.S.B.M. RI 5753.
22. Jacobson, M., Nagy, J. and Cooper, A. R. (1962). *Explosibility of dusts used in the plastics industry,* U.S.B.M. RI 5971.
23. Jacobson, M., Cooper, A. R. and Nagy, J. (1964). *Explosibility of metal powders,* U.S.B.M., RI 6516.
24. Hartmann, I. (1948). *Indus. Engg Chem.,* **40,** 752.
25. Kühnen, G. (1967). *Staub,* **27,** 529.
26. Smith, R. B. (1956). *Nucleonics,* **14,** (12), 28.
27. Dolan, J. E. and Dempster, P. B. (1955). *J. appl. Chem.,* **5,** 510.
28. Laffitte, P. and Bouchet, R. (1958). Suppression of explosion waves in gaseous mixtures by means of fine powders. *Proceedings of seventh international symposium on combustion.* Butterworth, London, 504.
29. *New Scientist* (1969), **43,** 192.

CHAPTER 9

CONCLUSION

There are two general themes involved in many of the applications of fine powders:

- —the extent of the surface of the powder;
- —the nature of the surface and its interaction with its environment.

The first of these requires little comment except to note that not all the surface may be available for interaction with any environment. This is especially true when much of the total surface exists in fairly fine pores or fissures which are too narrow to permit the easy entry of solids, or even liquids. Similarly, if the powder is used in a flowing gas stream, *e.g.* in gas chromatography or catalysis, penetration into fine pores is not likely to be complete and not all the surface will be available for reaction. In certain circumstances, therefore, powders prepared by condensation in which the particles themselves are non-porous may be preferable to those resulting from thermal decomposition which usually contain micropores.

In considering the present uses of fine powders it becomes plain that much more could be done to improve and control interaction with the matrix. At the moment the powders are used without much systematic modification. Oxides usually carry a multi-layer of adsorbed or combined water which may impede the formation of a good bond to a matrix. Such water may be very difficult to completely remove. Many fine oxides can be coated in such a way as to produce a hydrophobic surface which should prevent water bonding to the powder, but which may not itself provide bonding (other than physical) to the matrix.

It may be expected that future research will be aimed at studying the interaction of fine powders with matrices under a wide range of

143

conditions in order to produce different strengths of bonding at the interface.

It may be desirable to treat this problem as one in adhesion and to use the close environmental control known to be necessary in such studies. In this way more meaningful results will be obtained than in direct strength or other tests on, say, an elastomer filled with fine powder. However, it may be difficult to establish the influence of particle shape and porosity on the interaction with the matrix without more direct experiments.

Somewhat related studies on the effect of different types of surface on the forces between the particles of a fine powder also appear to be relevant.

The role played by hydrogen bonding between hydroxylated surfaces in complementing electrostatic and Van der Waal's forces may go some way towards explaining the very low relative densities of such powders.

Studies on these lines on anhydrous, coated and hydroxylated powders would be relevant to the handling, dispersion and compaction of fine powders.

Lastly, apart from carbon black, fine powders now used are all oxides. It is likely that the range of oxides used will be extended to include, for example, zirconia and ferric oxide and double oxides such as ferrites. Non-oxides, such as carbides and nitrides, are also likely to become available. Nothing is known about the surface chemistry of non-oxides or double oxides of most types, and it is likely that research will show that a very wide range of surface properties can be produced.

INDEX

Abrasion of elastomers, 91
Aerosols, preparation, 11
Agglomeration in surface coatings,
  111
Alloys
  dispersion-strengthened, 88
  fine, preparation by vapour phase
    reaction, 20
Alumina
  coating on titania, 114
  fine
    hot-pressed, water and CO_2 in,
      124
    hot pressing, 125
    preparation using
      centrifugal liquid wall fur-
        nace, 16
      high intensity arc, 13
      thermal decomposition, 26,
        28
      vapour phase reaction, 20
    top density, 137
  hot pressing, 125
  hydrophobic, 71
  isoelectric point in aqueous elec-
    trolyte, 75
  metastable phases, preparation,
    12
  silica-coated, 71
  strength, effect of surface, 127
  strengthened by molybdenum
    dispersion, 90
  substrate, as, 128
  transparency, 127
Atomisation, 5

Bound rubber, 94
Bridging between powder particles,
  135
Brunauer–Emmett–Teller (BET)
  gas adsorption isotherm,
  52

Carbides
  cermets, in, 100
  fine, preparation using
    high intensity arc, 13
    thermal decomposition, 30
    vapour phase reaction, 20
Carbon black
  bonding to elastomers, 94
  chaining, 93, 94
  dispersion, light scattering and
    adsorption, 110
  preparation, 21, 67
  production UK, 2
Centrifugal liquid wall furnace, 14
Centrifugal sedimentation, 46
Chalking of paint films, 114
Chemisorption
  effect on adsorption isotherms, 53
  strength of bonding in, 66
Coercivity of a fine grained microstructure, 128
Coulter Counter, 48
Critical nucleus, size of, 11

Dehydration leading to fine
  powders, 27
Dehydroxylation of silica, 68

145

Diazo coatings on powder, 117
Dispersion of
 fillers in elastomers, 93, 98
 hard phase in cermets, 100
 particles for electron microscopy, 47
 pigments in paint, stabilisation, 112

Elastomers, deterioration of, 91
Electron microscopy of fine powders, 37, 47
Electro-optical materials, 131
Electrophotographic paper, 117
Esterification of fine powder surfaces, 69, 71

Ferrite
 barium for permanent magnets, 129
 dielectric losses at high frequency, 130
 grain size for computer memories, 130
 nickel–zinc as low loss ferromagnet, 130
 fine, preparation, 24, 26, 29
Fine powder definition, 1
Flocculation, 22

Gas adsorption isotherms, 52
Grinding, 5, 6

Hiding power, 108
High intensity arc, 13
Hydrates, thermal decomposition of, 27
Hydrophobic oxides, 69, 71, 73
Hydroxylated surfaces, 68, 69, 71, 72
Hysteresis in gas adsorption isotherms, 59

Induction plasma, 17
Internal oxidation, 85, 87
Iron, fine, preparation, 13
Iron oxide
 fine, preparation, 13, 20, 26, 28
 pigment, as, 115

Kelvin equation, 57
Kubelka–Munk relationship, 108

Light absorption by pigments, 110
Liquid wall furnace, 14

Magnesia
 fine, preparation, 13, 16, 28, 31
 fully dense by hot pressing, 124
Metals, fine, preparation, 13, 20, 32
Metastable phases obtained by quenching, 12, 40
Micronisation, 7
Mullins softening of elastomers, 92

Nitrides, fine, preparation, 20
Non-porous particles formed by condensation, 43

Ostwald ripening, 21
Oxides
 double, fine, preparation by
  high intensity arc, 13
  precipitation, 24
  thermal decomposition, 26, 29
 fine, preparation by
  high intensity arc, 13
  liquid wall furnace, 16
  thermal decomposition, 26, 28
  vapour phase reaction, 20
 mixed, fine, preparation, 26, 29
 salts giving on thermal decomposition, 28, 31